Anonymus

Meteorological Stations and Observers of the Smithsonian

Institution in North America and Adjacent Islands

From the Year 1849 up to the Year 1869

Anonymus

Meteorological Stations and Observers of the Smithsonian Institution in North America and Adjacent Islands
From the Year 1849 up to the Year 1869

ISBN/EAN: 9783741180149

Manufactured in Europe, USA, Canada, Australia, Japa

Cover: Foto ©berggeist007 / pixelio.de

Manufactured and distributed by brebook publishing software
(www.brebook.com)

Anonymus

Meteorological Stations and Observers of the Smithsonian Institution in North America and Adjacent Islands

METEOROLOGICAL

IONS AND OBSERVE

OF THE

THSONIAN INSTITUTIO

IN

NORTH AMERICA AND ADJACENT I

FROM

THE YEAR 1846 UP TO THE END OF THE YEAR 18

WASHINGTON:

LIST

OF

SMITHSONIAN METEOROLOGICAL STATIONS AND OBSERVERS IN NORTH AMERICA AND ADJACENT ISLANDS FROM 1849 UP TO THE END OF THE YEAR 1868.

Those marked with a * are discontinued.

Name of station.	Name of observer.	Years of observations.	Name of station.	Name of observer.	Years of observations.
ANTILLES.			**HUDSON'S BAY TERRITORY.**		
Sombrero Island...	Alexis Julien..........	1863-'64			
Sombrero Island...	Milton Drayton......	1865	Abbitibbe post	Jas. Lockhart	1868
BAHAMAS.			Fort Anderson ..	R. Macfarlane	1863
			Fort George ...		1863
Turk's Island	J. B. Hayne..........	1859	Fort Liard	*R. Kennicott ..	1860
Turk's Island	J. C. Crimon, Capt. W. Hamilton.	1860	Fort Nascopee ...	K. Connolly.........	1863-'65
Turk's Island	*A. G. Carothers	1861	Fort Norman	Andrew Flatt.......	1861-'63
Turk's Island	United States consul.	1862-'65	Fort Rae...........	Lawrence Clark. Jr ..	1859-'60
Turk's Island	J. C. Crimon, United	1868	Fort Rae, Great Slave L.	Mrs. Lawrence Clarke.	1861-'64
Turk's Isl. Salt Cay	Samuel G. Garland.. States vice consul.	1861	Fort Simpson, Grt. Slave L.	B. R. Ross	1848-'61
Kanaam, N. P.......	A. M. Smith..........	1858-'59	Keenoghmisere	Thos. Richards......	1861-'63
BERMUDA.			Little Whale River	Walter Dickson	1862
			Moose Factory ,....	J. Mackenzie........	1857-'68
Hamilton	Capt. Alexander, R.E.	1852	Moose Factory to Lake Superior.	Colin Rankin.........	1862
Hamilton	Royal Gazette	1857	Red River settlem't.	Donald Gunn........	1857-'61
Whitby Bay	James B. Arnold	1857	Rigolet, Labrador	H. Connolly.........	1858-'60
St. George's	James Crawford	1858			
St. George's	Centre signal station, Royal Engineers, in Royal Gazette.	1856-'68	**JAMAICA.**		
			Upper Park Camp.	James G. Sawkins...	1855-'56
Ireland Island.....	John G. Calder	1859	Upper Park Camp.	Col. W. B. Marlow ..	1855
CANADA.			**MEXICO.**		
Clifton	W. Martin Jones, U. States consul.	1867-'68	Chinameca........	Charles Lassle	1859
Hamilton	Dr. W. Craigie	1857-'62	Cordova	J. A. Illote	1858-'57
Kingston..........	J. Williamson, (dir'r Kingston Observ'y.)	1856-'60			1862, '64
Michipicoten	Colin Rankin.......	1865-'66	Frontera, Tabasco.	Charles Lawle	1865
Montreal..........	Dr. A. Hall	1855-'63	Mexico	Prof. L. C. Ervendberg	1858-'56
Montreal..........	Thos. Blackwell	1861	Minititlan	Charles Lassle	1858-'60
Niagara	H. Phillips	1861-'63	Minador	Dr. Chas. Sartorius..	1858-'63
Port Neuf.........	Observations publis'd in the Naturaliste Canadien.	1868	San Juan Bautisto.	Charles Lassle	1861-'64
St. Martin's, near Montreal.	Dr. Chas. Smallwood.	1859-'63	Tuapas	Benjamin Crowther..	1867
Stanbridge	J. C. Baker	1857-'65	Vera Cruz	Hermann Berendt, M.D.	1850
Stanbridge	A. H. Gilmour	1868	**NEW BRUNSWICK.**		
Stratford..........	C. J. Macgregor	1863-'63			
Toronto	Magnetic Observatory	1855-'68	St. Johns	G. Murdock	1859-'60
Toronto	Capt. J. H. Lefroy ..	1849	**NEWFOUNDLAND.**		
		1851-'52	St. Johns	*Jas. Delany, Jr., and *E. M. J. Delany.	1857-'64
COSTA RICA.			St. Johns	H. D. M. Miln'y Post.	1849
Limon.............	Felipe Valentin	1865-'68	St. Johns	Rev. R. O. Caswell, M. A.	1868
San José..........	C. N. Riotte	1865-'66			
San José..........	Dr. A. Von Frantzius.	1856-'67	**NEW GRANADA.**		
San José..........	Oficina Central di Estadistica.	1867-'68	Aspinwall	W. T. White, M. D...	1857-'58
GUATEMALA.			Aspinwall	*J. P. Kinge, M. D...	1863-'68
Guatemala........	Antonio Cwndas.....	1856-'62	Aspinwall	*J. P. Kinge, M. D., and G. V. Hacker, M. D.	1864, 1867
HONDURAS.			**NICARAGUA.**		
Belize	S. Cockburn	1862-'68	....	J. Morre............	1851
Trujillo	E. Pardot	1854			

List of Smithsonian meteorological stations and observers—Continued.

Name of station.	Name of observer.	Years of observations.	Name of station.	Name of observer.	Years of observations.
NOVA SCOTIA.			**ALASKA.**		
Halifax	Royal Engineers	1858–'61	Fort Youkon	* R. Kennicott	1861
Halifax	Col. W. J. Myers, F. R. M.	1863–'65	Nulato	W. H. Dall	1866–'67
Halifax	Board of Trade	1854	Sitka	Dr. Alex. H. Hoff	1867–'68
Halifax	R. J. Nelson	1830	St. Michaels	H. M. Bannister	1865–'66
Horton	C. F. Hartt	1858	St. Michaels	J. M. Head	1865–'66
Pictou, (Albion Mines.)	Henry Poole	1843–'55	Unalaklect	F. Westdahl	1866–'67
Windsor	King's College	1857–'62	**ARIZONA.**		
Wolfville	Acadia College	1854	Fort Whipple	Dr. E. Coues, U. S. A	1863
Wolfville	Prof. D. F. Higgins	1867–'68	**ARKANSAS.**		
Wolfville	C. F. Hartt	1858	Arkadelphia	Dennis Barlow	1859
Wolfville	Prof. A. P. S. Stuart	1855–'59	Arkadelphia	Female College	1859
PORTO RICO.			Bentonville	Paul Graham	1859–'61
			Brownsville	B. F. Coulter	1859–'60
Est. San Ysidro	George Latimer	1868	Buckhorn	Armistead Younger	1859
SAN SALVADOR.			Dardsville	Silas Rea Helleth	1861
La Union	Charles Dorat, M. D.	1868	Fort Smith	Rev. Francis Springer	1866–'67
ST. DOMINGO.			Gainesville	James T. Banks	1859
	Jonathan Elliott	1860	Green Grove	Robert Burch, M. D.	1860
VANCOUVER'S ISLAND.			Helena	O. F. Russell	1863–'68
Victoria	David Walker, M. D	1863–'64	Jacksonport	O. Alexander Martin, M. D.	1859–'60
ALABAMA.			Little Rock	Philip L. Anthony	1849
Asheville	Thomas M. Barker	1857	Mico	Rev. H. F. Buckner	1860
Auburn	Prof. John Darby	1854–'59	Mountain Home	J. H. Howard	1861
Brown	Dr. Chas. F. Percival	1849–'51	Perryville	W. H. Blackwell	1859–'61
Boligee	Col. Horace Harding	1861	Perryville	H. F. Hardy	1856
Bon Secours	W. J. Van Kirk	1866–'67	Spring Hill	P. F. Finley	1859
Cahaba	Matthew Troy, M. D.	1859	Spring Hill	J. Reynolds	1859–'60
Carlowville	H. L. Allam, M. D	1858–'60, 1867–'68	Spring Hill	P. F. Finley and J. Reynolds	1860
Erie	Dr. Sam'l K. Jennings	1849	Waldron	Geo. W. Featherstone	1859–'60
Erie	Dr. T. C. Osborne	1851–'58	Washington	Alex. P. Moore, M. D	1861
Eutaw	A. Winchell	1851–'59	Washington	Dr. K. D. Smith	1849–'51
Fish River	W. J. Van Kirk	1868	Yellville	J. W. Wenel	1859–'60
Greensboro'	Robert B. Waller	1866–'69	Yellville	W. B. Flippin	1859–'60
Greensboro'	N. T. Lupton	1868	**CALIFORNIA.**		
Green Springs	H. Tutwiler	1854–'59	Auburn	Robert Gordon	1859–'60
Havana	Prof. H. Tutwiler	1853, 1858–'61, 1866–'68	Columbia	Silas Earle, M. D.	1857–'60
Havana, six miles east of.	R. K. Jennings, M. D.	1868	Crescent City	Robert B. Randall	1859–'60
Livingston	Rev. S. D. Smith	1859–'60	Downieville	T. R. Kibbe, B. D.	1860
McKeith's P. O.	R. T. Meriwether	1854	Folsom	Rev. R. V. Blakeslee	1861
Mobile	Dr. S. B. North	1849	Fort Yuma		1857
Mobile	Rev. J. J. Nicholson	1859	Hotcut	James Siever	1859
Monroeville	S. J. Cumming	1849, 1851–'55	Honcut	J. Slaven and Mrs. E. N. Dunham	1860
Montgomery	Rev. J. A. Shepherd	1859–'61	Honcut	Mrs. E. S. Dunham	1861–'63
Montgomery	W. L. Foster	1859–'60	Mare Island	U. S. Naval Hospital.	1868
Moulton	Andrew J. Harris	1868	Marsh's Ranche	Francis M. Rogers	1867–'68
Moulton	Thomas J. Peters	1868–'69	Maritime	Edwin Howe	1860
Moulton	Prof. J. Shackelford	1861	Marysville	W. C. Belcher	1857–'59
Moulton	Ashley D. Hurt	1859			1861–'63
Opelika	Dr. H. A. Shields	1867–'68	Meadow Valley	James H. Whitlock	1860–'68
Orville	Dr. S. K. Jennings	1859–'60	Meadow Valley	Colbert A. Canfield, M. D.	1864
Orville	T. A. Hinton and J. A. Coleman.	1860	Meadow Valley	M. D. Smith	1863–'64
Prairie Bluff	William Henderson	1857	Mokelumne Hill	Wesley K. Beecher	1859–'61
Prairie Bluff	B. M. Reynolds	1857	Monterey	Colbert A. Canfield, M. D.	1859–'60, 1864–'65
Selma	Dr. S. K. Jennings	1859–'59	Murphy's	Ephraim Carting	1858
Spring Hill	A. Corprite, S. J	1868	Presidio of San Francisco.	W. W. Hays, M. D.	1849
Tuscaloosa	Prof M. Tuomey	1853–'54	Presidio of San Francisco.	D. F. Parkinson	1863–'64
Tuscaloosa	George Benagh	1854–'55	Presidio of San Francisco.	Post surgeon	1859–'61
Union Springs	J. L. Mandrie	1868	Sacramento	Dr. F. W. Hatch	1864
Uniontown	Rev. K. A. Cobbs	1859–'60	Sacramento	Drs. F. W. Hatch and T. M. Logan.	1853
Wetobaville	Benj. F. Holly	1849, 1851–'54	Sacramento	Dr. T. M. Logan	1849–'55
			Sacramento	Charles Craft	1863
			San Francisco	W. O. Ayres, M. D.	1854–'55, 1865–'68
			San Francisco	Dr. H. Gibbons	1854–'55

List of Smithsonian meteorological stations and observers.—Continued.

Name of station	Name of observer	Years of observations	Name of station	Name of observer	Years of observations
CALIFORNIA—Con.			**DIST. OF COLUMBIA.**		
Santa Barbara	W. W. Hays, M. D. ...	1864	Georgetown	Rev. C. H. McKee ...	1860-'63
Santa Clara	Prof. O. S. Frömbers..	1859-'61	Washington	U. S. N. Observatory	1858-'60
Santa Clara	Lewis A. Gould......	1859			1861-'67
Spanish Rancho ...	M. D. Smith	1862-'63	Washington	J. Wismer........	1857-'59
Spanish Rancho ...	Mrs. M. D. Smith ...	1864-'66			
Stockton	Dr. Robert K. Bird ..	1855-'56	**FLORIDA.**		
Stockton	Walter L. Trivett...	1857			
Union Rancho.....	W. L. Dunham	1858	Alligator	Edward R. Ives.....	1857-'58
			Atsena Otis	Hon. Aug. Steele ...	1856-'61
COLORADO.			Belair..........	Benj. F. Whitney ...	1857-'58
			Cedar Keys	Judge Aug. Steele...	1851-'58
Central City......	W. D. McLain	1860-'61	Cedar Keys	W. G. Andrus	1857
Denver City......	D. C. Collier	1859	Chemot Hill	John Newton......	1851
Denver City......	Fred. J. Stanton ...	1862	Fernandina	* Henry M. Corey...	1857
Fountain.........	Arthur N. Merriam ..	1866-'67	Gainesville......	James H. Linley ...	1855-'61
Golden City......	E. L. Berthond.....	1867	Gomon..........	P. C. Garvin, M. D...	1866
Montgomery	James Luttrell.....	1862-'63	Gordon.........	H. B. Scott........	1864-'69
Mountain City....	Dr. William T. Ellis.	1860-'62	Green Cove Spring	O. A. Boardman	1862
			Hibernia........	F. L. Barbdler.....	1857-'58
CONNECTICUT.			Jacksonville.....	Dr. A. S. Baldwin...	1853-'60
					1864-'69
Brookfield........	Sanford W. Roe.....	1868	Key West (Salt	W. C. Dennis......	1854-'64
Canton..........	Jarvis Case........	1861-'63	Pond.)		
Colebrook	Miss C. Rockwell ...	1860-'62	Key West (Mag-	George D. Allen	1860-'61
Columbia	W. G. Yeomans....	1862-'64	netic Observ'y.)		
East Windsor Hill.	P. A. Chadbourne ...	1852	Key West (Mag-	O. P. Ferguson, and	1861-'64
Georgetown......	Aaron B. Hull......	1855-'57	netic Observ'y,)	J. G. Oltmans,	
Groton..........	Rev. E. Dewhurst...	1866-'68	Knox Hill.......	John Newton......	1856-'59
Hartford........	Charles H. Hundley.	1849-'51	Lake City.......	Edward H. Ives....	1858-'60
Middletown	Prof. J. Johnston ...	1854, '58			1865-'68
		1858-'59	Lake City.......	Gabe M. Fisher.....	1867
Middletown	* Prof. A. W. Smith..	1849, '53	Lake City.......	Rev. W. W. Keep ...	1864
		1862	Micanopy	Dr. James B. Bean...	1864-'60
New Haven......	H. G. Du Bois, Jr ...	1859	Orange Mills	John Newton......	1864
New Haven......	D. C. Lewvroworth..	1862-'64	Pensacola.......	U. S. Navy Yard....	1849
New Haven......	Prof. E. Casler	1849, '51	Pensacola.......	J. Pearson, U. S. N..	1851-'58
New London.....	Rev. Tryon Edwards,	1849, '51	Pensacola.......	J. Pearson and Lieut.	1853
	D. D.	1855-'59		Joseph Fry,	
North Colebrook ..	M. H. Cobb	1849, '51	Pensacola.......	Lieut. Joseph Fry ...	1856
Norwich.........	N. Scholfield......	1855-'56	Pensacola.......	Lieuts. Jos. Fry and	1857
Plymouth	Dwight W. Learned .	1862-'64		J. W. Hewitt.	
Pomfret.........	Rev. Daniel Hunt...	1863-'68	Pensacola.......	Lieut. J. W. Hewitt..	1858
Salisbury........	Dr. Ovid Plumb.....	1849, '51	Port Orange.....	J. M. Hawks, M. D...	1867-'69
		1852-'54	St. Augustine....	Dr. John E. Peck....	1849
Saybrook	James Rankin	1859-'61	St. Augustine....	P. S. Manran, M. D..	1854-'60
Wallingford.....	Benj. F. Harrison ...	1856-'62	Tallahassee.....	Benj. P. Whitner ...	1859-'61
Waterbury......	Rev. B. G. Williams.	1857-'68	Tallahassee.....	W. S. Bogert......	1852
West Cornwall....	T. S. Gold........	1854	Tallahassee.....	Lardner Gibbon.....	1859-'60
Windsor	R. H. Phelps......	1851	Uchee Anna.....	John Newton......	1849
			Warrington.....	Thayer Abert......	1858-'60
DAKOTA.					
			GEORGIA.		
Fort Union......	F. G. Riter........	1857-'59			
Greenwood.......	Freeman Norvell...	1859-'61	Athens..........	Prof. John D. Easter	1857-'59
Yankton	M. K. Armstrong ...	1865	Atlanta.........	J. G. Westmoreland	1859-'60
Yankton.........	(G. D. Hill, G. W. Law-	1862	Atlanta.........	Fred. Dreimer & Son	1865-'66
	son, H. G. Williams.		Augusta.........	William Haines....	1854-'57
Yankton	H. G. Williams.....	1863	Augusta.........	William Schley....	1854
			Augusta.........	Wm. H. Doughty, M.D	1858-'60
DELAWARE.			Boston.........	Rev. W. Hewitt....	1865-'66
			Clarkesville.....	Jarvis Van Buren ...	1859-'67
Delaware City....	L. Venkels	1866-'67	Clarkesville.....	Col. J. R. Stanford..	1858
Dover..........	J. P. Walker......	1854	Covington......	Benjamin F. Camp ..	1859-'61
Georgetown......	Dr. D. W. Maull....	1859	Colloden........	John Darby.......	1852-'53
Lewes	John Barton.......	1849	Cuthbert.......	Charles C. Reavey ..	1866
Milford	R. A. Martin.......	1857-'59	Dalton.........	J. R. McAfee......	1861
Newark.........	Prof. W. A. Norton..	1849	Darien.........	Charles Grant	1849
Newark.........	Prof. E. D. Porter ..	1852	Factory Mills	F. T. Simpson.....	1857
Newark.........	Prof. W. A. Crawford	1854	Hillsboro'.......	Eli S. Glover......	1857-'58
Newark.........	Prof. W. A. Crawford,	1855	Macon..........	Miss L. J. Whitney ..	1866
	R. A. Martin.		Macon..........	John A. Rockwell ..	1862
Newark.........	Prof. W. A. Crawford,	1856	Macon..........	J. F. Adams.......	1860
	R. A. Marcn, and T.		Madison........	Prof. Wm. D. Williams	1854
	J. Craven.		Milledgeville....	J. M. Cotting.....	1849
Newark.........	Thos. J. Craven and	1857	Milledgeville....	Prof. C. W. Lane...	1849
	Mrs. E. D. Porter.		Penfield.......	Prof. J. E. Willet...	1850
Newark.........	Mrs. E. D. Porter ...	1858	Perry..........	Dr. Geo. P. Cooper .	1851-'59
Newark.........	Robert Crawford ...	1856	Pasomath......	James M. Reed	1857
Wilmington.....	* Urban D. Hedges...	1862-'63			

List of Smithsonian meteorological stations and observers.—Continued.

Name of station.	Name of observer.	Years of observations.	Name of station.	Name of observer.	Years of observations.
GEORGIA—Con'd			**ILLINOIS—Con'd**		
Powellton	P. C. Pendleton	1852	Evanston	* W. R. Morrison	1864
Savannah	* Dr. John F. Posey	1852-'59	Evanston	H. W. Scoville	1864
Savannah	R. T. Gibson	1859-'61	Evanston	Jos. H. Gill and others	1865
Sparta	Dr. E. M. Pendleton	1852-'61	Evanston	Fred. J. Hipe	1866-'67
Summerville	Stephen Elliott Burnham	1859	Evanston	Prof. Oliver Marcy	1862
The Rock	Dr. James Anderson	1853-'60	Farmbridge	Elmer Baldwin	1860
Thomasville	Rev. W. Blewitt	1860	Fremont Centre	Isaac H. Smith	1872-'76
Thomson	Dr. W. T. Grant	1852-'59	Oakwan	Emil Howell	1859-'60
Wilmington Island	K. T. Gibson	1859-'58	Galesburg	Prof. Wm. Livingston	1861-'66
Zebulon	Mrs. J. T. Arnold	1857-'59	Golconda	Rev. Wm. V. Eldridge	1862-'63
			Granville	L. O. Edgerly	1851
IDAHO.			Granville	J. L. Jenkins	1857
Cantonment Jordan	W. W. Johnson	1859	Hazel Dell	Henry Griffing	1853-'55
Fort Benton	M. C. Bowman	1863	Ironsepia	Smiley Sheppard	1861
Fort Halleck	J. H. Pinfrock	1864	Highland	A. F. Bandelier, jr	1860-'61
Fort Laramie	Col. W. O. Collins	1863-'64	Hillsboro	John K. Tilroush	1858
Fort Laramie	A. F. Ziegler, M. D.	1865	Hoylton	J. Ellsworth	1864-'65
			Hoylton	G. J. March	1868
ILLINOIS			Jacksonville	Prof. Wm. Coffin	1868
			Jacksonville	Timothy Dudley	1861-'64
Albany	Warren Olds	1861-'62	Lacon	A. H. Thompson	1867
Albion	Edgar P. Thompson	1857	Lebanon	Prof. N. E. Cobleigh	1860-'61
Alton	M. Y. McMasters	1849	Lomami	Timothy Dudley	1867-'68
Alton	Norton Johnson	1849	Magnolia	Henry K. Smith	1866-'69
Assiniboia	E. H. Bowman, M. D.	1866-'69	Manchester	John Grant	1854-'61
Athens	Joel Hall	1851-'5?	Manchester	John Grant and Miss Ellen Grant	1862-'63
Augusta	Dr. N. B. Mead	1849-'50	Manchester	John & C. W. Grant	1865-'66
		1851-'66	Marengo	O. P. Rogers	1868
Aurora	* Andrew J. Babcock	1857-'61	Marengo	O. P. & J. S. Rogers	1868-'69
Aurora	Abram Spaulding	1863-'66			1863-'66
Batavia	Prof. Wm. Coffin	1853-'54	Marengo	F. Rogers	1860
Batavia	* Thomp'n Mead, M. D	1857-'61	Monroe	Silas Meacham	1869
Batavia	E. Coyen	1858-'59	Mount Sterling	Rev. Alex. Duncan	1865-'68
Batavia	Frank Crandon	1861-'62	Naperville	Lewis Ellsworth	1859
Belleville	N. T. Baker	1860-'63	Naperville	Milton N. Ellsworth	1859-'60
Belleville	Dr. John J. Patrick	1861	Newton	Rev. Wm. V. Eldridge	1860
Belleville	Dr. J. J. Patrick and N. T. Baker	1862	North Prairie	C. H. Bryant	1866
Belvidere	G. H. Moss	1866	Olney	Rev. H. M. Belchertein	1860
Bloomington	Jesse Allison	1850-'61	Oneida	J. K. Pashley, M. D.	1869-'71
Brighton	Wm. V. Eldridge	1875-'58	Ottawa	Dr. J. O. Harris	1852-'61
Carbon Cliff	Mrs. Wm. H. Thomas	1850	Ottawa	Geo. O. Smith, M. D.	1858-'60
Carthage	Samuel J. Wallace	1857	Ottawa	Samuel L. Shotwell	1862
Carthage	Mrs. E. M. A. Bell & Sam'l J. Wallace	1859	Ottawa	Mrs. Emily H. Merwin	1863-'69
Centralia	H. A. Kelmoher	1863	Paris	C. Lving	1868
Channahon	Rev. D. H. Sherman	1860	Pekin	J. H. Bishet	1857-'65
Channahon	Dr. Jos. Pitch	1861	Peoria	Dr. Fred. Brendel	1855-'66
Chicago	Henry Talcott	1851	Peoria	M. A. Breed	1861-'63
Chicago	G. D. Hiscox	1850-'77	Plymouth	Dr. J. H. N. Klliger	1854
Chicago	Samuel Brookes	1859-'59	Quincy	Rev. O. D. Giddings	1849
Chicago	M. C. Armstrong and J. H. Roe	1860-'61	Ridge Farm, Vermillion county.	B. C. Williams	1864
Chicago	Gustave A. Boettner	1860-'81	Riley	E. Babcock	1866-'67
		1861	Robinson's Mills	I. Brendel, M. D	1864
Chicago	A. M. Byrne, J. H. Roe, and others.	1862	Rochelle, (Alta)	Daniel Cary	1866-'68
Chicago	John O'Donoghoe	1862	Rockford	William Holt	1849
Chicago	Arthur M. Byrne	1863-'64	Sandwich	K. E. Hallon, M. D	1859-'08
Chicago	Isaac A. Pool	1866	South Pass	Frank Baker	1857-'58
Chicago	John G. Langguth, jr	1867-'68	South Pass	H. C. Spaulding	1861-'66
Clinton	C. H. Moore	1864-'66	Sheath Pass	H. C. Freeman	1867-'68
De Kalb	John D. Parker	1866	Springfield	Geo. W. Brinkerhoff	1865-'66
Dixon	J. Thos. Little	1856-'63	Tiskilwa	Verry Aldrich	1858-'59
		1867	Upper Alton	Prof. P. P. Brown	1851-'59
Dongola	Ralph C. Meeker	1861-'62	Upper Alton	* Dr. John James	1859-'59
Dubois	Wm. C. Spencer	1865-'66	Upper Alton	Anan James	1859-'59
Du Quoin	C. Ziegler	1864	Upper Alton	Mrs. Anna C. Tribla	1861-'64
Edgar county	J. W. Bowen	1858	Wapella	T. Lewis Groff	1864
Edgington	Dr. E. H. Bowman	1857-'6?	Warsaw	Benj. Whitaker	1853-'59
Elgin	Jas. B. Newcomb	1856-'61	Waterloo	H. Kanater	1865
Elsah	Orestes A. Blanchard	1862-'63	Waukegan	Dr. Wm. Jodyn	1849

List of Smithsonian meteorological stations and observers—Continued.

Name of station.	Name of observer.	Years of observations.	Name of station.	Name of observer.	Years of observations.
ILLINOIS—Con'd.			INDIANA—Con'd.		
Willow Creek	E. C. Bacon	1859–63	Rockville	M. H. Anderson	1859–'66
Willow Hill	Henry Griffing	1863	Rockville	J. W. Tenbrook	1859
Winnebago Depot	J. W. Tolman	1859–'60	Shelbyville	J. T. Bullock	1859–'62
Woodstock	Geo. R. Bassett	1859–'61	South Bend	Prof. Gardner Jones	1851
Wyanet	E. S. Phelps	1864	South Bend. (Notre	Prof. Thos. Vagnier	1858–'59
Wyanet	E. S. Phelps and Miss	1865–'66	Dame Univer'y.)		
	L. E. Phelps.		South Bend	Miss G. Webb	1859
York Neck	V. P. Gay	1864–'65	South Bend	James H. Dayton	1860–'61
			South Bend	Renben Barrough	1851–'63
INDIANA.			South Hanover	Prof. S. H. Thomson	1849
			Spiceland	Wm. Dawson	1851–'56
Aurora	Geo. Sutton, M. D.	1839	Vevay	Charles G. Boerner	1861–'63
		1865–'66	Walnut Hills	W. W. Austin	1849
Balbec	Miriam Griest	1865–'66			
Bloomingdale	Wm. D. Hobbs	1864	INDIAN TERRITORY.		
Bloomingdale	Miss M. A. Hobbs	1865			
Bloomington	Prof. C. M. Dodd,	1868	Armstrong Acad'y	Prof. A. G. Moffat	1849
	assisted by T. H.		Doaksville	P. F. Brown	1849
	Mallow and others.		Talequah	T. B. Van Horne	1859
Cadiz	Wm. Dawson	1854–'63			
Cannelton	Hamilton Smith, jr	1856–'61	IOWA.		
Carthage	Charles M. Hobbs	1864			
Columbia	Dr. F. McCoy and	1865–'66	Algona	F. McCoy, M. D.	1860
	Miss Lizzie McCoy.		Algona	F. McCoy and Miss	1861–'63
Evansville	John P. Crisp	1857–'59		Elizabeth McCoy.	
Fort Wayne	Prof. A. C. Huestis	1849	Algona	Philip Dorweiler	1863–'64
Fort Wayne	Miss G. Webb	1860–'61	Algona	James H. Warren	1867–'68
Greencastle	Prof. Jos. Tingley	1851–'54	Athens	B. Carpenter	1847
Greencastle	Wm. H. Larrabee	1859–'63	Bangor	Isaac M. Gilley	1871–'83
Indianapolis	* Royal Mayhew	1861–'63	Bellevue	John C. Ferry	1856–'60
Indianapolis	W. W. Butterfield	1864–'65	Boonsboro	C. Babcock	1867–'69
Indianapolis	W. W. Butterfield	1866–67	Border Plains	G. C. and W. K. Gaat	1858
	and Mrs. Butterfield.		Border Plains	Wm. K. Gaat	1857–'59
Indianapolis	W. J. Elstun	1867–'68	Bowen's Prairie	Samuel Woodworth	1868
Jalapa	Abner G. Irwin	1848	Burlington	John M. Corse	1850–'57
Kendallville	W. B. Coventry	1854	Burlington	Louisa P. Love	1868–'69
Kendallville	J. Kanner	1854	Burlington	Mrs. James Love	1868
Knightstown	D. Dean	1868	Ceres	John M. Lingenalek	1865–'68
Lafayette	A. H. Bixby	1854	Clarinda	S. H. Kridelbaugh,	1865–'66
Lafayette	H. Peters	1834		M. D.	
Lafayette	Isaac E. Windle	1865	Clinton	Nathan R. Parker	1856–'58
Laporte	B. M. Newkirk	1849	Clinton	P. J. Farnsworth	1866–'68
Leo	W. W. Spratt, M. D.	1861	Dakota	Wm. O. Atkinson	1867–'68
Logansport	Charles B. Lasolle	1857–'28	Davenport	Nathan H. Parker	1858
Logansport	Isaac Bartlett	1858–'61	Davenport	A. J. Pinley	1858
Logansport	Thos. B. Helm	1863	Davenport	H. S. Finley	1859
Madison	C. Barnes	1854	Davenport	H. S. Finley and W.	1860
Madison	Rev. Samuel Collins	1864–'69		P. Danwoody.	
Madison	Oliver Malvey	1865	Davenport	J. Chamberlain, W.	1861
Marion	Thomas Holmes	1866–'68		P. Danwoody, H.	
Michigan City	C. S. Woodard	1857–'58		H. Belfield.	
Michigan City	W. Woodbridge, S.	1859–'60	Davenport	Dr. Ignatius Langes	1861
	D. Angell, and H.		Davenport	H. H. Belfield and W.	1862
	Rinke.			P. Danwoody.	
Milton	Dr. V. Kersey	1853–'55	Davenport	J. Chamberlain and	1863
Mishawaka	Geo. C. Mansfield	1859		W. P. Danwoody.	
Monett	E. J. Rice	1863–'64	Davenport	J. Chamberlain	1864
Muncie	G. W. H. Kemper	1866–'68	Davenport	Geo. B. Pratt	1865
New Albany	C. Barnes	1855–'56	Davenport	G. B. Pratt and Syd-	1866
New Albany	Dr. Alex. Martin	1859		ney Smith.	
New Albany	Dr. E. S. Crozier	1863–'65	Davenport	D. C. Sheldon	1867–'69
New Castle	Prof. Jos. Tingley	1849	Des Moines	Rev. J. A. Nash	1855–'57
New Castle	Thos. B. Redding	1865–'65	Dubuque	Dr. Asa Horr	1851–'55
New Garden	D. H. Roberts	1854			1857–'65
New Harmony	John Chappelsmith	1852–'68	Dubuque, (Alexander College.)	Rev. Joshua Phelps	1854
New Harmony	Dr. D. D. Owen	1849–'51			
		1867	Dubuque	Dr. W. W. Woolsey	1858
Newport	Daniel H. Roberts	1851	Fairbanks	Dexter Beal	1858
Paoli	A. P. Turner	1859	Fairfield	J. M. Shaffer	1856–'60
Poseville	John Griest	1864	Fairfield	Miss Sue McBeth	1859
Remaster	J. H. Longbridge, M. D	1864–'65	Fayette	John M. McKenzie	1859–'60
		1867–'68	Fontanelle	A. P. Bryant	1865–'66
Richmond	Dr. Jno. T. Plummer	1849, '51	Forestville	Daniel Sheldon	1859–'63
Richmond	W. W. Austin	1851–'55	Fort Madison	Daniel McCready	1852, '54
		1858–'61			1865–'66
Richmond	Joseph Moore	1855–'59	Franklin	Dexter Beal and W.	1857
Richmond	John Haines	1859–'60			

List of Smithsonian meteorological stations and observers—Continued.

Name of station.	Name of observer.	Years of observations.	Name of station.	Name of observer.	Years of observations.
IOWA—Continued.			**KANSAS—Con'd.**		
Fort Dodge	C. N. Jorgenson	1867-'68	Fort Riley	J. M. Shaffer and E. P. Camp.	1868
Grove Hill	Dexter Beal	1859-'60			
Grove Hill	Dexter Beal and W. W. Beal.	1861	Gardner	O. F. Merriam	1860
Grove Hill	Mrs. Celia Beal	1862	Gardner	Joseph Scott	1861-'62
Guttenburg	Philip Dorweiler	1864-'65	Holton	Dr. James Walters	1867
Guttenburg	James P. Dickinson	1865-'68	Junction City	E. W. Seymour, M. D.	1868
Harris Grove	Jacob F. Stern	1866-'68	Lawrence	G. W. Brown	1857-'59
Hesper	H. R. Williams	1860-'61	Lawrence	W. J. R. Blackman	1860-'61
Independence	D. S. Derring	1861-'67	Lawrence	A. N. Fuller	1862-'64
Independence	Alex. Camp Wheaton	1862-'66	Lawrence	W. L. G. Soule	1863-'64
Independence	Mrs. D. D. Wheaton	1866-'68	Lawrence	Geo. W. Hollingworth	1867
Independence	Geo. Warne, M. D.	1867-'68	Lawrence	Prof. F. H. Snow	1868
Iowa City	Hermann H. Falrall	1859	Leavenworth	H. D. McCarty	1857-'59
Iowa City	W. Reynolds	1857-'58			1868
Iowa City	Prof. Theo. N. Parvin	1861-'62	Leavenworth	E. L. Berthoud	1859-'60
Iowa Falls	Nathan Townsend	1862-'64	Leavenworth	M. Shaw	1860-'62
Keokuk	Miss Ida E. Ball	1853-'54	Leavenworth	Dr. J. Stayman	1865-'62
Keokuk	Dr. J. E. Ball	1853	Leavenworth	T. B. Stowell	1866
Keokuk	Prof. K. M. Taylor	1858	Lecompton	Dr. Wm. T. Ellis	1859-'60
Kossuth	Wm. F. Leonard	1862	Lecompton	Wm. A. McCormick	1860-'61
Kossuth	Isaiah Reed	1860-'61	Lecompton	David G. Bacon	1866
Lyons	A. T. Hudson, M. D	1858-'67	Leroy	J. G. Shoemaker	1867
Lyons	P. J. Farnsworth	1862-'65	Manhattan	Isaac T. Goodnow	1857-'62
Lyons	Dr. J. Newman	1866	Manhattan	Rev. N. O. Preston	1858-'60
Manchester	* Allen Mead	1865-'68	Manhattan	I. T. Goodnow and H. L. Denton.	1863
Maquoketa	Edward F. Hobart	1857			
Marble Rock	H. Wadey	1867-'68	Manhattan	Henry L. Denton	1864
Monticello	Chauncey Mead	1864-'66	Manhattan	Agricultural College. (R. F. Mudge, and and others.)	1863-'68
Monticello	M. M. Moulton	1866-'68			
Mount Pleasant	E. L. Briggs	1863-'64	Mapleton	S. O. Kimoe, M. D.	1857-'59
Mount Vernon	Prof. R. Wilson Smith	1857	Moneka	J. O. Wattles and Celestia Wattles.	1858
Mount Vernon	Prof. Alonzo Collins	1860-'62			
Muscatine	T. S. Parvin	1849-'50	Neosho Falls	R. F. Goss	1858-'59
		1851-'52	Neosho Falls	Mrs. K. W. Greenbush	1860
		1853-'59	Olatha	W. Brokwith	1864-'68
Muscatine	P. G. Parvin	1853-'54	Ridgeway	O. H. Brown	1863
Muscatine	Rod Foster	1860-'64	Topeka	F. W. Giles	1858
Muscatine	T. S. Parvin and Rev. John Ufford.	1860	Wyandot	John H. Millar	1859-'60
Muscatine	Rev. John Ufford	1861-'62	**KENTUCKY.**		
Muscatine	Josiah P. Walton	1863-'68			
Onawa	Richard Bicbbins	1864	Ballardsville	Dr. John Swain	1853-'54
Osage	Rev. Alva Bush	1866-'67			1860-'69
Pella	E. H. A. Scheeper	1854-'56	Bardstown	*John H. Labmann	1856
Pleasant Plain	Townsend McConnell	1855-'65	Bardstown	J. H. Labemann and Thos. H. Miles.	1859
Pleasant Spring	Rev. B. F. Odell	1856			
Plum Spring	B. F. Odell and Miss Mary G. Odell.	1855	Bardstown	Thos. H. Miles	1860-'61
			Beech Fork	Dr. C. D. Cuse	1860
Plum Spring	Rev. B. F. Odell	1859	Bowling Green	J. K. Youngiere	1849-'50
Poultney	Dr. B. F. Odell	1853-'54			1851-'59
Quinnquaton	Dr. E. C. Bidwell	1853-'56	Bowling Green	F. C. Henrich	1852
Rolfe	Oscar L. Strong	1868	Cliftonburg	Dr. Samuel D. Martin	1865-'68
Rossville	Carlisle D. Brannan	1867-'58	Clinton	Rev. T. H. Cleland	1860
Sioux City	Dr. J. J. Saville	1867-'68	Danville	G. Beatty	1853-'58
Sioux City	A. J. Millard	1871-'68			1865-'68
St. Mary's	D. E. Read	1853	Danville	R. H. Caldwell	1864
Vernon Springs	Gregory Marshall	1861-'63	Drennon Springs	Prof. S. V. McMasters	1851
Washington	C. R. Royle	1861	Hardinsburg	Mrs. Mary A. Walker and J. C. Barbage.	1859
Waterloo	L. H. Doyle	1858-'64			
Waterloo	T. Steed	1864-'65	Hardinsburg	Joshua C. Barbage	1860-'61
Whitesboro	David M. Witter	1867-'68	Lexington	J. D. Shane	1854
			Lexington	Rev. N. R. Williams	1859
KANSAS.					1867-'68
Atchison	Dr. H. B. Horn and Miss Clotilde Horn.	1865-'68	Leaton	W. S. Denk	1865-'69
			Louisville	Rev. S. R. Williams	1858-'59
Avon	Allen Crocker	1868	Louisville	H. N. Woodruff	1862-'63
Baxter Springs	Ingraham & Hyland	1867-'68	Maysville	E. L. Berthoud	1853-'54
Burlingame	* Lucian Fish	1859-'61	Millersburg	Rev. J. Miller	1853
Cayuga	Wm. H. Gilmore	1859	Millersburg	Rev. J. Miller, Rev. G. S. Savage.	1854
Colesville	Rev. J. H. Drummond	1858-'59			
Council City	Edmund Fish	1857-'58	Millersburg	Dr. Geo. S. Savage	1855-'68
Council Grove	A. Woodworth, M. D	1863-'68	Newport	Prof. M. G. Williams	1861
Emporia	C. F. Oakfield	1862	Nicholasville	Jos. Meb. Matthews, D. D.	1861-'63
Fort Riley	Rev. David Clarkson	1858-'60			
Fort Riley	Dr. Fred. P. Drew, U. S. A.	1860-'64	Nolin	J. Grinnell	1858
			Paducah	Andrew Mattson	1858-'59
			Paris	L. G. Ray	1854-'55

List of Smithsonian meteorological stations and observers—Continued.

Name of station.	Name of observer.	Years of observations.	Name of station.	Name of observer.	Years of observations.
KENTUCKY—Con'd.			**MAINE—Con'd.**		
Prospect Hill	O. Beatty	1849-'61	Oldtown	Rev. S. H. Merrill	1849-'50
Haselville	E. M. March	1850			1851-'53
Springdale, (near Louisville.)	Mrs. L. Young	1849-'50	Oxford	Howard D. Smith	1862
		1851-'55	Paris	H. Everett	1849
		1857-'60	Penobscot	Rev. E. Dewhurst	1862
Taylorsville	H. C. Mathis	1866	Perry	William B. Dusu	1852-'65
			Portland	Henry Willis	1853-'60
LOUISIANA.			Portland	John W. Adams	1850-'61
			Rumford Point	Waldo Pettingill	1865-'66
Breton	J. H. Carter	1867-'68	Setoc	Edwin Thomas	1864
Falls River	A. W. Jackson, M. D.	1858	South Thomaston	Joshua Bartlett	1853-'54
Grand Coteau	D. F. Anthonioz	1860	Standish	John P. Monkton	1843-'48
Independence	Col. C. H. Swasey	1859	Steuben	J. D. Parker	1849-'50
Independence	Mrs. M. J. Maukard	1860			1851-'60
Jackson	Prof. W. P. Riddell	1854	Thomaston	George Prince and	1849-'50
New Orleans	Dr. E. H. Barton	1849, '50		Chr. Prince.	1851-'52
		1851-'57	Topsham	Warren Johnson	1852-'61
New Orleans	Lewis B. Taylor	1856-'57	Vassalboro	James Van Blarcom	1852-'63
		1858, '61	Warren	Calvin Bickford	1848-'49
New Orleans	Dr. S. P. Moore, U.S.A.	1860	Webster	Almon Robinson	1845-'47
			West Waterville	B. F. Wilbar	1853-'64
New Orleans	Harrison Thompson	1861	Whitehead	Joshua Bartlett	1849-'50
New Orleans	Robert W. Foster	1867-'69			1851-'52
New Orleans	E. L. Rankett	1868	Williamsburg	Edwin Pitman	1853, '68
St. Francisville	B. R. Gifford	1858	Williamsburg	E. Pitman	1867-'68
Trinity	A. K. Kilpatrick, M. D.	1858-'59	Wyndham	Samuel A. Everett	1849-'70
Trinity	Edward Merrill, M. D.	1858-'56			1851-'56
		1860	**MARYLAND.**		
Vidalia plantation	Rev. A. K. Teale	1867	Agricult'l College, Prince George Co.	Montg. Johns, M. D.	1861-'62
MAINE.			Annapolis	Prof. W. P. Hopkins	1851
			Annapolis	A. Zambreek, M. D.	1853-'56
Bangor	Stephen Gilman	1849	Annapolis	W. H. Goodman	1856-'60
Bangor	C. L. Nichols	1850-'60	Baltimore	Dr. Lewis F. Steiner	1852-'59
Belfast	G. Emerson Brackett	1850-'64	Baltimore	Prof. Alfred M. Mayer	1857-'59
Bethel	Rev. A. G. Gaines	1851-'62	Bladensburg	Benj. O. Lowndes	1854-'64
Biddeford	J. G. Garland	1849-'50	Catonsville	George S. Grape	1863-'67
		1851-'53	Chestertown	James A. Pearce, jr.	1853-'57
Biddeford	F. A. Small	1854	Charlestown	Prof. A. W. Clark	1858
Isle Hill	Rev. S. H. Merrill	1854-'55	Charlestown	Rev. A. Sutton	1859-'60
Isle Hill	H. H. Osgood	1866	Charlestown	Prof. J. Russell Dutton	1861-'64
Brunswick	Prof. Parker Cleaveland	1849-'59	Cumberland	T. C. Atkinson	1849
Bucksport	Rufus Buck	1849-'50	Ellicott's Mills	Philip Tabb	1864
		1851-'56	Emmittsburg	J. H. Smith	1846-'50
Carmel	J. J. Bell	1853-'57	Emmittsburg	Prof. C. H. Jourdan	1847-'48
Castine	Dr. J. L. Stevens	1851	Frederick	Dr. Lewis F. Steiner	1851
Cornish	G. W. Gapill	1855-'68	Frederick	Henry E. Hamshew	1853-'54
Corinth	Silas Weed	1857-'58			1856-'63
Dexter	A. F. Wilbar	1861-'63	Frederick	Miss H. M. Baer	1863-'64
East Exeter	Stephen Gilman	1854	Hagerstown	Rev. J. P. Carter	1852-'54
East Wilton	Henry Reynolds, and Lauriston Reynolds.	1861-'63	Lebanon	Lewis J. Bell	1852
Exeter	Dr. J. B. Wilson	1860-'61	Leitersburg	Jacob E. Bell	1858-'62
Foxcroft	M. Pitman	1863-'64	Leonardtown	Dr. Alex. McWilliams	1856-'59
Freedom	E. A. Usher	1859	New Windsor	Prof. J. P. Nelson	1852
Fryeburg	G. H. Barrows	1849-'58	New Windsor	Prof. J. F. Maguire	1854
Gardiner	Hon. R. H. Gardiner	1855-'64	Nottingham	A. P. Dalrymple	1852
Gardiner	Rev. F. Gardiner	1864	Oakland	L. B. Cofran	1852-'54
Gardiner	Rev. F. and R. H. Gardiner	1865	Port Deposit	Henry W. Thorp	1849
Gardiner	R. H. Gardiner	1866-'68	Ridge	T. O. Knight	1856-'57
Maryland	E. E. Brown, R. W. Hall, L. A. Strickland, and others.	1859	Sandy Hill	Isaac Bond	1849
			Sykesville	Prof. William Baer	1849-'50
Hiram	Prin. Wadsworth	1849-'84			1851-'52
Henlson	Hiram Weich	1849			1853-'54
Levee	Benj. M. Towle	1866-'67	Sykesville	Prof. Wm. Baer and Miss H. M. Baer.	1853-'54
Levee	E. Pitman	1863-'64	Sykesville	Miss H. M. Baer	1871-'75
Livingston	W. G. Lord	1849-'84	St. Inigoes	Rev. Jas. Stephenson	1854-'84
Lisbon	Ann P. Moore	1859-'60	Union Bridge	Warrington Gilling-ham.	1864
Monson	B. F. Wilbar	1864-'59			
Newcastle	C. L. Nichols	1859	Walkersville	Josiah Jones	1849-'51
New Sharon	J. F. Pratt, M. D.	1862-'69	Woodlawn	James O. McCormick	1863-'64
North Brigade	A. H. Wyman	1859-'60	**MASSACHUSETTS.**		
North Bridgton	M. Gould	1852-'61	Amherst	Prof. E. S. Snell	1849-'69
North Prospect	Virgil G. Eaton	1867	Baldwinsville	Rev. E. Dewhurst	1863-'65
Norway	G. W. Verrill, jr.	1856-'61	Barnstable	M. B. Gifford	1852-'55

List of Smithsonian meteorological stations and observers—Continued.

Name of station.	Name of observer.	Years of observations.	Name of station.	Name of observer.	Years of observations.
MASS.—Continued.			**MASS.—Continued.**		
Boston	E. L. Smith	1857	Worcester	Drs. Ed. A. Smith, F. H. Nies, and others.	1852–'56
Boston	E. L. Adams	1858			1845
Bridgewater	Marshal Conant	1854	Worcester	Dr. Geo. Chandler	1854
Bridgewater	L. A. Darling	1856–'57	Worcester	John N. Hargent and others.	1857–'56
Bridgewater	C. W. Felt and others.	1856–'59			
Bridgewater	Normal School	1862–'64	Worcester	Dr. H. C. Prentiss	1859–'64
Brookline	Rev. John D. Perry	1859	Worcester	Joseph Draper, M. D.	1865–'68
Byfield	Martin N. Root	1854			
Cambridge	W. C. Bond	1855–'58	**MICHIGAN.**		
Cambridge	Harvard College Observatory.	1858–'60			
Cambridge	Augustus Fendler	1865–'66	Alpena	J. W. Paxton	1865–'68
Canton	B. H. Ellis	1857–'58	Ann Arbor	Dr. H. R. Schetterly	1852
Chelsea	Naval Hospital	1851–'54	Ann Arbor	L. Woodruff	1852–'54
Clinton	Geo. M. Morse, M. D.	1850–'61			1856–'57
Danvers	A. W. Mack	1858–'59	Ann Arbor	Prof. A. Winchell	1854
Duxbury	James Ritchie	1849			1856–'57
Fall River	Charles C. Terry	1851	Ann Arbor	L. Woodruff and A. Winchell.	1855
Fitchburg	George Raymond	1860–'61			
Florida	L. F. Whitcomb	1857–'64	Battle Creek	Dr. W. M. Campbell	1849–'69
Framingham	Gustavus A. Hyde	1849	Birch	Dr. Thos. Wright-y	1848–'54
Georgetown	Henry M. Brison	1865–'67	Brooklyn	Dr. M. K. Taylor	1858–'54
Georgetown	S. Augustus Nelson	1857–'58	Burr Oak	Charles Betts	1849–'50
Grafton	Rev. Wm. G. Scandlin	1859–'61	Central Mine	S. B. Whittlesey	1857–'60
Hinsdale	Rev. E. Newhurst	1858	Clifton	Wm. Van Orden, jr.	1860–'61
Kingston	Guilford S. Newcomb	1858–'59	Clinton	Eleazer Wainwright	1851–'59
Lawrence	John Fallos	1857–'59	Coldwater	X. C. Southworth	1849
Lowell	Charles J. Gillis	1849–'51	Cooper	Mrs. Octavia C. Walker.	1854–'58
Lunenburg	Geo. A. Cunningham	1858–'59			1859–'63
Lynn	Jacob Batchelder	1849–'50	Copper Falls	Chas. K. Whitney	1856–'57
		1851–'52	Corunna	Hebet Crane	1853
Mendon	Henry Rice	1849	Detroit	Wm. A. Raymond	1849
Mendon	Mr. John G. Metcalf	1849–'59	Detroit	Rev. Geo. Duffield	1852–'58
Milton	Rev. A. K. Teele	1862–'64	Detroit	Dr. Zena Pitcher and J. M. Norton.	1858–'60
Nantucket	Hon. Wm. Mitchell	1858–'61			
New Bedford	Thomas Bailey	1849–'51	Detroit	U. S. Engineers	1860–'63
New Bedford	Samuel Hodman	1858–'60	Detroit	Dr. Zena Pitcher	1861–'62
New Bedford	Edward T. Tucker	1856–'57	Eagle River	Mrs. M. A. Goff	1858
Newbury	John H. Caldwell	1865–'66	East Saginaw	Dr. S. F. Mitchell	1854
Newburyport	Dr. B. C. Perkins	1855–'58	Flint	Dr. D. Clark	1854–'58
North Attleboro'	Henry Rice	1851–'54	Forestville	Lieut. C. N. Turnbull	1854
North Billerica	Rev. Silas Nason	1865–'68	Fort Gratiot	Lieut. C. N. Turnbull, U. S. A.	1858–'59
Plainfield	Francis Shaw	1857			
Princeton	Hon. John Brooks	1853–'57	Garden	Edwin Ellis, M. D.	1864
Randolph	Orrin A. Reynolds	1861–'62	Grand Haven	Heber Squier	1850–'63
Richmond	William Bacon	1849–'50	Grand Rapids	Franklin Everett	1849
		1851–'52	Grand Rapids	Dr. J. Hollister	1849–'51
		1851–'53	Grand Rapids	Alfred O. Currier	1854–'58
		1859–'60	Grand Rapids	L. H. Streng	1857–'60
Rockport	R. D. Muncy	1854	Grand Rapids	Edwin A. Strong	1860–'61
Roxbury	Benjamin Kent	1849	Grand Rapids	J. R. Parker	1864
Sandwich	N. Barrows, M. D.	1862–'65	Grand Rapids	E. S. Holmes	1863–'68
South Groton	Alfred Collis	1859	Grand Traverse	H. R. Schetterly	1854
Southwick	Aurora Holcomb	1849–'57	Holland	L. H. Streng	1860–'61
Springfield	Lucius C. Allis	1853–'59			1863–'68
Springfield	Francis A. Brewer	1859	Homestead	George E. Steele	1861–'67
Stockbridge	Abraham S. Peet	1849	Houghton	J. B. Minick	1863–'68
Taunton	Albert Schlegel	1854–'57	Howell	Dr. H. M. Schetterly	1849–'50
Topsfield	Nathan W. Brown	1852–'53			1851–'52
Topsfield	John H. Caldwell	1853–'64	Kalamazoo	Harmon M. Smith	1864–'67
Topsfield	Arthur M. Merriam	1864–'66	Kalamazoo	Milton Chase	1863–'67
Topsfield	Sidney A. Merriam	1866–'68	Kalamazoo	Frank Little	1858
Uxbridge	Dr. James Robbins	1854	Lake George	J. H. Foster and Edward Perrault.	1859
West Dennis	Eugene Tappan	1854			1859
Westfield	Rev. Dr. E. Davis	1854–'65	Lansing	Cleveland Abbe	1859
West Newton	John H. Bixby	1857–'58	Lansing	J. C. Holmes	1859
Weymouth	Dr. N. Q. Tirrell	1856–'57	Lansing	Prof. R. C. Kedzie	1863–'73
		1858	Litchfield	R. Bullard	1863–'68
Williamstown	C. M. Freeman	1851–'59	Lower Saginaw	James G. Birney	1849
Williamsburg	Prof. P. A. Chadbourne	1851	Manchester	F. M. Reamer, M. D.	1864
Williamstown	D. J. Holan, jun. Orton, Lavallette Wilson, and others.	1864–'57	Marquette	Peter White	1857
			Marquette	Dr. O. H. Blaker, jr.	1858–'61
Williamstown	I. Metler, C. J. Lyons, M. L. Borger & others.	1857–'59	Marquette	and P. M. Mason.	1862–'63
Williamstown	Astronomical Observatory.	1859–'60	Mill Point	Rev. L. M. S. Smith	1861–'65
Williamstown	Prof. Albert Hopkins	1860	Monroe	Thomas Whelpley	1850
Wood's Hole	R. R. Gifford	1854–'55	Monroe	Capt. A. B. Perkins	1854
Worcester	S. F. Haven	1849–'52	Monroe	Miss H. J. Whelpley	1855–'60
			Monroe	O. W. Rowlsky	1858–'61

List of Smithsonian meteorological stations and observers—Continued

Name of station.	Name of observer.	Years of observations.	Name of station.	Name of observer.	Years of observations.
MICHIGAN—Con'd.			**MINNESOTA—Co'd.**		
Monroe	Miss Helen 1, and Florence Whelpley.	1851	Smithfield	B. C. Livings	1849
			Stillwater	A. Van Voorhies	1856
Monroe	Miss F. C. Whelpley.	1863–'68	St. Anthony's Falls.	C. F. Anderson	1854
Monroe Piers	John Lane	1859–'63	St. Cloud	O. E. Garrison	1861–'62
Muskegon	H. A. Pattison	1868	St. Joseph's	Rev. D. D. Spencer	1853–'55
New Buffalo	J. B. Crosby	1852–'59	St. Joseph's	A. O. Kollam	1854
Northport	H. R. Scherterley	1862–'63	St. Paul	Rev. A. B. Paterson	1866–'67
Northport	Rev. Geo. N. Smith	1863–'68		D. D.	
Ontonagon	H. Selby	1851–'63	St. Paul	John W. Helmstreet	1866–'67
Ontonagon	Edwin Ellis, M. D.	1865–'68		hury.	
Ottawa Point	John Oliver	1859–'61	Tamarack	Mary A. Grave	1863–'64
Otsego	Matthew Coffin	1850–'52	Travers des Sioux	Rev. R. Hopkins	1849–'51
Otsego	Milton Chase	1851	Winnebaw	Spencer L. Hillier	1857–'58
Oshtemo & elsew'o.	Henry H. Mayes	1864–'69			
Pennsylvania Mine.	Richard H. Griffith	1868	**MISSISSIPPI.**		
Pleasanton	Joseph D. Millard	1868			
Pontiac	James A. Weeks	1864	Brook Haven	T. J. D. Kernan	1867–'68
Port Huron	James Allen, Jr	1857–'59	Columbus	James N. Lull	1855–'59
Port Huron	Geo. A. Stockwell	1860	Como	C. W. Brekwith	1868
Bedford Centre	Chas. C. Smith, M. D.	1861	Fayette	Rev. T. H. Cleland	1866–'67
Romeo	Isaac Stone	1853	Gainesville	Charles A. Folsom	1849
Romeo	Seth L. and O. P. An-	1858	Clarksdaleville	Rev. E. S. Robinson	1853–'55
	drews.		Graysville	James H. Vincent	1849
Romeo	R. L. Andrews, M. D.	1855, '57	Grenada	Wm. Henry Waddell	1854
Saugatuck	L. H. Strong	1854–'56	Grenada	Prof. Albert Moore	1858–'60
St. James	James J. Strong	1853–'56			1866–'68
Sugar Island	U. S. Engineers	1863	Hernando	Wm. M. Johnston	1858–'60
Tawas City	U. S. Engineers	1861–'63	Jackson	Thomas Oakley	1853–'58
Thunder Bay	J. L. Malden	1859–'63	Jackson	A. R. Green	1854
Ypsilanti	Miss O. Webb	1850	Kingston	J. Edward Smith	1866–'67
Ypsilanti	C. S. Woodward	1859–'61	Marion	T. W. Plorer, M. D.	1849
			McLeod's	David Moore	1849
MINNESOTA.			Monticello	J. R. Cribbs	1860–'61
			Natchez	Geo. L. C. Davis	1849–'51
Afton	Dr. B. F. Babcock	1863–'67	Natchez	J. Edward Smith	1850
Beaver Bay	Thomas Clark	1858–'59	Natchez	R. McCary	1854–'61
Beaver Bay	Henry Wieland	1859–'60			1864–'66
Beaver Bay	Thos. Clark and C.	1860	Natchez	W. McCary	1868–'64
	Wieland.		Oxford	Prof. L. Harper	1851–'59
Beaver Bay	C. Wieland	1861–'69	Pass Christian	Rev. J. A. Shepperd	1869
Bowles Creek	Andrew Stouffer	1863–'66	Paulding	Rev. E. S. Robinson	1850–'59
Buchanan	Stephen Walsh	1857–'58	Port Gibson	Prof. J. Lloyd Elliott	1855–'57
Burlington	A. A. Hibbard	1859–'60	Prairie Line	Rev. E. S. Robinson	1850–'61
Cass Lake	Alonzo Barnard	1852	Vicksburg	A. L. Hatch	1849–'52
Cass Lake Mission	Rev. N. F. Odell	1856	Westville	J. H. Cribbs	1858–'60
Chatfield	T. F. Thickston	1859–'61	Yazoo City	Col. C. B. Sweezy	1860–'61
Danville	Thomas A. Kellett	1858			
Fond du Lac	Rev. Joseph W. Holt	1850–'51	**MISSOURI.**		
Forest City	A. C. Smith	1850–'61			
Forest City	Henry L. Smith	1862–'66	Allenton	Aug. Fendler	1865–'69
Fort Ripley	Rev. N. W. Manney	1864	Athens	John T. Caldwell	1861–'68
Grand Portage	Richard Borden	1857	Augusta	Conrad Mullenbrodt	1859
Hastings	T. F. Thickston	1861–'62	Bethany	D. J. Heaston	1859–'60
Hazelwood	S. R. Riggs	1855–'58	Bolivar	W. J. Vankirk	1858–'61
Hennepin county	J. B. Clough	1864–'65	Bolivar	James A. Rare	1868
Itasca	O. H. Kelley	1860–'67	Boonerville	Noris Sutherland	1850–'61
		1869	Canton	George P. Hoy	1861–'68
Lac qui Parle	Rev. S. L. Riggs	1852–'53	Canton	Dr. J. N. Parker	1868
Lac qui Parle	S. R. and A. L. Riggs	1854	Cape Girardeau	Rev. James Knoud	1–56–'58
Lake Winnibigosh-	Rev. Benj. F. Odell	1858	Carrollton	John Campbell	1859
ish.			Carrollton	R. J. Hufsaker	1859
Lapham	E. M. Wright	1857	Carrollton	D. J. Kirby	1862
Lapham	J. F. McMullen and	1858	Caseville	M. L. Wyrich	1859–'61

List of Smithsonian meteorological stations and observers—Continued.

Name of station.	Name of observer.	Years of observations.	Name of station.	Name of observer.	Years of observations.
MISSOURI—Cont'd.			**NEBRASKA—Con'd.**		
Hermitage	Miss Belle Moore	1867-'68	Omaha	James P. Allan	1860-'61
Harrisonville	W. H. Horner	1858-'61	Omaha	C. R. Wells	1860
Jefferson City	Nicolas De Wyl	1860	Peru	J. M. McKenzie	1867
Kirksville	Robert Byers, M. D.	1859	Rock Bluffs	H. C. Pardon	1860-'61
Kirksville	J. H. Myers	1859-'63	South Pass wagon road expedition	C. H. Miller	1859-'60
Laborville	William Muir	1863-'64			
Lancaster	John M. Weatherford	1859	**NEVADA.**		
Lexington	Joseph A. Wilson	1859			
Lexington	Geo. W. Wilson, jr	1860	Star City	R. C. Johnson	1865
Lexington	P. A. Wilson	1861			
Luray	B. P. Hogan	1860-'61	**NEW HAMPSHIRE.**		
Oregon	William Koecher	1867-'68			
Paris	W. F. Mazry	1859-'62	Antrim	Rev. Wm. Harris	1868-'69
Rhineland	Charles Vogel	1859-'60	Claremont	P. N. Freeman	1859-'50
Richmond	R. W. Finley	1859-'60	Claremont	Arthur Chase	1859-'68
Rockport	C. Q. Chamber, M. D.	1855-'58	Claremont	Stephen O. Mead	1864-'67
Rolla	Homer Ruggles	1867-'68	Claremont	Linus Stevens	1867-'69
Springfield	J. A. Stephens	1857-'59	Concord	Dr. Wm. Prescott	1856-'57
St. Joseph	Edward B. Kreig	1859-'60	Concord	H. E. Sawyer	1857-'58
St. Louis	Dr. Geo. Engelmann	1855-'57	Concord	E. P. Colby	1859
St. Louis	A. Wislizenus, M. D	1856-'57	Concord	John T. Wheeler	1859-'62
St. Louis	G. Engelmann, M. D., and A. Wislizenus, M. D.	1858	Concord	James C. Knox	1860
			Dublin	Rev. L. W. Leonard	1859 1861-'58
St. Louis	Augustus Fendler	1859-'64	Dunbarton	Alfred Colby	1868
St. Louis	J. H. Leuermann	1860-'63 1864	Exeter	Rev. L. W. Leonard	1833-'55
St. Louis	Rev. P. W. Koning	1861	Exeter	Rev. Elias Nason	1861-'65
St. Louis	Rev. F. H. Stuntebeck	1863-'68	Farmington	Louis Bell	1860-'61
St. Louis	Rev. I. Meyerinanus	1868	Francestown	Dr. Martin N. Root	1857
Stockton	Wm. Wells	1860-'61	Francestown	A. H. Bixby	1859-'62
Toronto	B. D. Dodson	1859-'60	Great Falls	Henry E. Sawyer	1856-'57
Trenton	Thomas J. Conkling	1859	Hanover	Prof. Ira Young and A. A. Young	1853-'54
Tuscumbia	Wm. M. Lamphin	1858			
Union	Dr. W. Moore	1864	Isle of Shoals	Thos. B. Laighton	1849
Union	Miss Belle Moore	1867	Laconia	J. W. French, agt. L.	1857-'61
Warrensburg	Rev. J. P. Pollock	1868	Lake Village	W. G. & W. M. Co	
Warrensburg	Marion F. Hamaker	1858	Littleton	Robert C. Whiting	1865-'61
Warrenton	Mary A. Tidswell	1860-'63	Londonderry	Robert C. Mack	1854-'57
Waynesville	R. O. Lingow	1858	London Ridge	Isaac S. French, M. D.	1862-'65
Westport	Rev. N. Scarritt	1851	Manchester	Hon. N. N. Hall	1855-'57 1859-'61
			North Barnstead	R. F. Hanscom	1857-'59
MONTANA.			North Deansted	Charles B. Plumm	1859-'60
Benton City	Dr. H. M. Lehman	1878	North Littleton	Rufus Smith	1858-'60
Camp Cooke	Dr. H. M. Lehman	1867	Portsmouth	Dr. C. Chase, U.S.N.	1849
Cantonment Wright	T. Kalcski	1861-'62	Portsmouth	John Hatch	1867-'68
Helena City	Alex. Camp Wheaton	1865-'68	Salmon Falls	George B. Sawyer	1853-'54 1855
			Shelburne	Fletcher Odell	1858-'68
NEBRASKA.			Stratford	B. Gould Brown	1855-'58
Bellevue	D. E. Reed	1854	Sarsham	Andrew Wiggins	1859-'68
Bellevue	Rev. Wm. Hamilton	1857-'57	Stratford	Branch Brown	1859-'68
Bellevue	Henry M. Bart	1857	Tamworth	Alfred Brewster	1857
Bellevue	Mrs. L. E. Caldwell	1868	Top of Mt. Washington	Joseph H. Huff	1859
Blackbird Hills	Rev. Wm. Hamilton	1857-'58	Wentworth	Peter L. Hoyt	1859
Brownville	Charles B. Smith	1858-'60	West Enfield	Nath. Parmari	1856-'59
Dakota City	H. B. Brown	1867-'68			
Deer Creek	Major Thos. S. Twiss	1858	**NEW JERSEY.**		
De Soto	Charles Sells	1867-'68			
Elkhorn City	Anza M. J. Howrd	1858-'64	Belleville	Thos. B. Merrick	1849
Elkhorn City	John S. Howrd	1858-'62	Bloomfield	B. L. Cooke	1848-'58 1862-'63
Fontanelle	John Evans	1858 1859-'63	Burlington	Prof. Adolph Frost	1849-'54
Fontanelle	Henry Gibson	1868	Burlington	Dr. E. R. Schmidt	1857
Fort Pierre	M. C. Brown	1850-'51	Burlington	Prof. A. Frost and Dr. E. R. Schmidt	1857-'58 1858
Fort Union	E. T. Denig	1854			
Glendale	A. L. Child, M. D.	1861 1866-'67	Burlington	John C. Denuvo	1852-'69
Glendale	Dr. A. C. Child and Mrs. J. C. Child.	1868	Cinnaminson	William Parry	1858-'60
			Cole's Landing	James R. Lippincott	1864-'65
Joala	L. J. Hill	1863	Dover	Howard Shriver	1866-'68
Kenosha	Rela White	1859-'62	Elwood	J. S. Fritts	1867-'69
Nebraska City	Edgar E. Mann	1859	Freehold	D. V. Simpson and D. S. Wills	1857-'58
Nebraska City	P. Zaborr	1859			
Norway Hill	R. O. Thompson	1865	Freehold	O. R. Wills	1858-'59
Omaha	Wm. N. Byers	1857-'59	Greenwich	Benj. Sheppard	1850-'51
Omaha	John O. Hale	1859-'60			

List of Smithsonian meteorological stations and observers—Continued.

Name of station.	Name of observer.	Years of observations.	Name of station.	Name of observer.	Years of observations.
N. JERSEY—Con'd.			**N. YORK—Con'd.**		
Greenwich	Clarkson Sheppard	1864	Clyde	Matthew Mackie	1859-'63
Greenwich	C. Sheppard and Miss R. C. Sheppard.	1865-'67	Constableville	I. L. Fairchild	1862
			Constantia	Stephen Clark	1861
Greenwich	Miss R. C. Sheppard	1868	Danville	Rev. John J. Brown	1850-'51
Haddonfield	John Clement, jr	1858	Depauville	Henry Haas	1865-'68
Haddonfield	Samuel Wood	1865-'68	East Franklin	Dr. J. W. Smith	1854
Lambertville	Jacob N. Gary	1869	East Henrietta	A. S. Wadsworth	1851-'52
Long Branch	Howard A. Stokes	1861, '63	Eden	Stephen Landon	1855
Long Branch	Arch. Alexander	1865	Edre	Anson S. Landon	1857-'59
Morristown	Dr. S. C. Thornton	1849, '61	Falconer	Laurens A. Langdon	1853-'54
Morristown	Miss E. E. Thornton	1859	Farmer	A. B. Covert	1859
Morristown	Thos. J. Beans	1865-'68	Farmingdale	John C. Merritt	1848
Morristown	Jos. W. Lippincott	1865	Fishkill Landing	W. H. Denning	1865-'68
Mount Holly	Morgan J. Rhees, M.D	1861-'68	Flatbush	Rev. Thos. H. Strong	1854-'55
Newark	W. A. Whitehead	1849-'60	Flatbush	Rev. R. D. Van Kleek	1856-'60
New Brunswick	Prof. Geo. H. Cook	1854	Flatbush	Rev. W. W. Howard	1860
		1863-'68	Flatbush	Rev. L. T. Murk	1858-'60
New Brunswick	E. T. Mack	1854	Fordham	John Ambler	1856
New Brunswick	Edwin Allen	1859	Fordham	Claudius Permot	1856-'57
New Brunswick	Edwin Allen and G. W. Thompson.	1860	Fordham	H. M. Paine, M. D.	1858
New Brunswick	Geo. W. Thompson	1861-'65	Fordham	Rev. Jno. Ambler and Prof. A. T. Monroe.	1858-'60
Newfield	L. D. Conch	1867-'68	Fort Ann	P. A. McMorn	1853-'55
New Germantown	Arthur B. Noll	1868	Fort Edward	Prof. Solomon Rice	1857-'59
Newton	Thos. Ryerson, M. D	1858	Fort Niagara	L. Leffman	1858-'63
Passaic Valley	Wm. Brooks	1862-'65	Fredonia	Pres. D. J. Pratt	1858, '63
Paterson	Wm. Brooks	1866-'68			1864
Progress	Thos. J. Beans	1862-'65	Friendship	George W. Frice	1864-'67
Readington	John Fleming	1866-'67	Garrisons	Thos. B. Arden	1860-'61
Riceville	Prof. L. Harper	1860-'61			1862-'63
Rio Grande	Jerusha R. Palmer	1869	Geneva	Rev. W. D. Wilson	1855-'57
Salem	C. M. Dodd	1856			1864-'68
Salem	George Watson	1859	Geneva	Job Ellerton	1859
Seaville	Barker Cole	1865-'67	Germantown	Wm. Tompkins	1859
Seaville	R. C. Cole	1869	Germantown	Rev. Sanford W. Roe	1856-'58
Sergeantville	John T. Sergeant	1857-'58	Governeur	Dr. P. O. Williams	1852-'54
Trenton	Ephraim M. Cook	1865-'68	Gouverneur	Cyrus H. Russell	1862-'63
Vineland	Jno. Ingram, M. D	1867-'68	Glen's Falls	Warren P. Adams	1855
Woodstown	George Watson	1860	Great Valley	Kathalo Kelsey	1858-'60
			Havana	Col. E. C. Frost	1859-'60
NEW MEXICO.			Hector	David Trowbridge	1865-'67
Pope's Expedition	James M. Rando	1855-'57	Hermitage	A. A. Hibbard	1860-'68
			Homer	Edwin O. Reed	1855-'57
NEW YORK.			Honeoville	Walter D. Yale	1849-'51
					1856-'60
Adam's Centre	C. D. Potter, M. D	1859-'61			1865-'68
Albany	H. M. Paine, M. D	1865-'66	Inst. for Deaf and Dumb, N. Y.	Prof. Orin W. Morris	1860-'68
Albion	L. F. Munger	1849-'54	Ilion	J. D. Ingersoll	1859-'63
Alps	James H. Ball	1849-'51	Jamestown	Rev. Sanford W. Roe	1863-'68
Angelica	E. M. Alba	1854-'58	Jericho, L. I	Albert G. Carll	1849
Auburn	John B. Dill	1860-'65	Lake	Peter Ried	1855-'58
Baldwinsville	John Bowman	1849-'57	Leroy	L. P. Munger	1854
Beaver Brook	C. R. Woodward	1853-'54	Leyden	C. G. Merriam	1858
Dellport	H. W. Tice	1857-'62	Liberty	John Fell	1855-'56
Beverley	Thos. B. Arden	1853-'59	Lima	Prof. N. A. Lattimore	1861
Blackwell's Island, N. Y.	W. W. Sanger, M. D	1853-'57	Little Genesee	Daniel Edwards	1860-'66
Brookhaven	E. A. Smith and daughters.	1859	Lockport	E. Giddings	1849
			Lockport	James B. Trevor	1849-'50
Buffalo	A. Hesmer	1849-'52	Lodi	John Lefferts	1849-'55
Buffalo	Ebenezer O. Salisbury	1853-'54	Lowville	Irah R. Adams	1854
		1860	Lowville	J. Caroll Mosner	1854-'58
Buffalo	Dr. S. B. Hunt	1854	Lyons	Dr. E. W. Sylvester	1856-'59
Buffalo	W. D. Allen	1854	Madrid	E. A. Dayton	1849-'59
Buffalo	William Ives	1858-'59	Marathon	Lewis Swift	1863
		1866-'68	McGrawville	J. Metcalf Smith	1855-'57
Buffalo	U. S. Engineers	1860-'63	Mexico	John M. French	1855-'57
Canton	E. W. Johnson	1853-'59	Minaville	D. R. Banning and J. W. Banning.	1867
Cazenovia	Prof. Aaron White	1853-'64			
Cazenovia	Prof. Wm. Neale	1863, '67	Minaville	J. W. Banning	1869
		1868	Mohawk	James Lewis	1861-'68
Charlotte	Andrew Mulligan	1850-'53	Morristown	William Day	1859
Charlton	Cornelius Chase	1849-'51	Moriches	E. A. Smith and Miss N. Smith.	1854-'67
Chatham	C. Thornton Chase	1853-'54			
Clinton	Prof. O. Root	1856	Morley	Ezra Parmelee	1849
Clinton	H. M. Paine, M. D	1857-'59	Newark Valley	Rev. Samuel Johnson	1868
		1862-'65	Newburg	James M. Gardiner	1864-'68
Clockville	J. P. Chapman	1848 &c.	New York	U. S. Naval Station	1849
			New York	J. S. Gibbons	1854

List of Smithsonian meteorological stations and observers—Continued.

Name of station.	Name of observer.	Years of observations.	Name of station.	Name of observer.	Years of observations.
N. YORK—Con'd.			**N. YORK—Con'd.**		
New York	S. De Witt Bloodgood	1854-'55	Spencertown	Levi S. Packard	1841
New York, (Eastern Dispensary.)	Caleb Swann and Dr. J. P. Loines	1854-'81	Springville	J. W. Fowle	1849
			Springville	Moss Lane	1851
New York	Fred. I. Klade	1863-'61	Stapleton	Spencer L. Hillier	1847-'68
New York, (Rutherford's Observatory.)	Charles O. Wakely	1860-'63	Suffern	James H. Warren	1862
			Syracuse	Henry L. Dinsmore	1851-'59
			Thorne	R. O. Gregory	1861-'68
New York	Naval hospital	1863-'68	Throg's Neck	Francis M. Rogers	1854, '68
New York, (N. Y. Skating Club.)	E. B. Cook	1863-'68	Throg's Neck	Miss Elizabeth Morris	1863-'68
			Troy	John W. Helmstreet	1849-'68
New York	Rev. John M. Aubier	1863-'67	Troy	Prof. E. A. H. Allen	1833-'54
New York, (Columbia College.)	Prof. Chas. A. Joy	1865-'69	Troy	Prof. Darrow Greene	1856-'57
			Troy	Wm. L. Haskin	1860-'61
New York, (Central Park Observatory.)	Madox Patrick Smith	1867-'68	Utica	Dr. L. A. Tourtellot	1856-'57
					1862
North Argyle	Geo. M. Hunt	1864	Utica	Joseph Graham	1860
North Hammond	Charles A. Wooster	1865-'68	Vermillion	N. H. Bartlett	1860-'68
North Salem	John F. Jenkins	1849-'53	Wales	S. O. Carpenter	1854
North Salem	Mrs. M. J. Lobdell	1855-'56	Wampsville	Dr. Stillman Spooner	1862-'63
North Volney	J. M. Parrick	1849	Warsaw	J. P. Morse	1863
Nichols	R. Howell	1857-'68	Waterburgh	David Trowbridge	1868
Ogdensburg	W. E. Guest	1849-'50	Waterford	John C. Henne	1856-'63
		1854-'60	Watertown	Dr. P. O. Williams	1853-'57
Oneida	Dr. Stillman Spooner	1864-'68	Waterville	James M. Tower	1849-'51
Oswego	C. Strong	1849	Wellsville	H. M. Shaver	1857-'58
Oswego	J. H. Hart	1851-'54			1861
Oswego	Capt. W. S. Malcolm	1854-'68			
Otis	Prof. Weston Flint	1861	West Cromwell	Lewis Woodward	1856-'57
Ovid	J. W. Chickering	1855-'59	West Day	Jude M. Young	1856-'59
Palmyra	W. S. Gilman, jr	1869	West Farms	J. S. Gorton	1856-'57
Palmyra	Stephen Hyde	1864-'65	West Morrisania	I. Zappful	1853-'59
Peekskill	Charles A. Lee	1854	White Plains	O. B. Willis	1862-'63
Penn Yan	Dr. H. P. Sartwell	1854-'57	Wilson	R. S. Holmes	1852-'64
		1859			
Perry City	David Trowbridge	1864	**NORTH CAROLINA.**		
Philipstown	Thos. B. Arden	1851-'59	Asheville	W. W. McDowell	1857-'58
Pine Hill	Godfrey Zimmerman	1859-'60	Asheville	E. J. Aston	1867-'68
Pittsville	J. D. Norton	1856-'57	Asheville	J. P. E. Hardy, M. D.	1868
Plattsburg	Joseph W. Taylor	1853-'59	Autaway Hill	F. J. Kean	1849-'62
Pompey Hill	S. Marshall Ingalls	1857-'58			1867-'69
Pompey Hill	John P. Kendall	1858	Chapel Hill	Prof. James Phillips	1859-'61
Poughkeepsie	Prof. C. H. Wuring	1849	Davidson College	Prof. W. C. Kerr	1859-'60
Rochester	Prof. Wetherill	1849	Gaston	Geo. F. Moore, M. D.	1856-'58
Rochester	* Prof. C. Dewey	1855-'67	Goldsborough	Prof. D. Morelle	1855-'59
Rochester	* Prof. M. M. Matthews	1858-'67	Greensboro'	Geo. P. Moore, M. D.	1850-'51
Rochester	H. Wells Mathews	1861	Goldsborough	Prof. E. W. Adams	1860-'61
Rochester	W. M. L. Fish	1868			1868
Rockett's Harbor	U. S. Naval Station	1849	Green Plains	Sam'l W. Westbrook	1850
Rockett's Harbor	Medsrin Lines	1851-'58	Guilford Mine	Alexander Wray	1867
Rockett's Harbor	H. Metcalf	1859-'61	Jackson	Rev. Fred. Fitzgerald	1858-'54
Sag Harbor	E. N. Byram	1849-'58	Kenansville	Prof. N. B. Webster	1858
Saratoga	Walter H. Biker	1856-'50	Lake Scuppernong	Rev. J. A. Sheppard	1849-'52
Rangerville	R. G. Williams	1863-'66	Lake Scuppernong	D. Morrell	1851
Rangerville	Jas. W. Grush, Jas. M. Alexander, and Levi S. Packard	1858-'69	Lincolnton	Dr. J. Bryant Smith	1854
			Newbern	Robert H. Drysdale	1858
			Murfreesboro'	Rev. A. McDowell	1856-'61
Schenectady	Robert M. Fullerand Haven V. Swart	1864	Oxford	John H. Mills	1868-'67
Schenectady	Akola A. Julien and H. A. Schanber.	1858-'59	Oxford	Wm. R. Hicks, M. D.	1867-'68
			Raleigh	T. Carter and W. H. Hamilton.	1858
Seneca Falls	Elisha Foote	1849	Raleigh	W. H. Hamilton	1849
Seneca Falls	John P. Fairchild	1849-'59	Raleigh	Rev. Fisk P. Brewer	1866-'68
Seneca Falls	Chas. A. Avery	1853-'54	Rutherfordtown	J. W. Calloway	1849
Seneca Falls	Pallo Cowing	1864-'64	Statesville	Thos. A. Allison	1858-'68
Seneca	Henry B. Fellows	1857	Thornbury	Rev. F. Fitzgerald	1854
Sherburne	Rev. Jas. H. Haswell	1865	Thornbury	Dan. Morelle	1854
Sing Sing	C. F. Maurice	1849-'52	Trinity College	Rev. B. Craven	1857-'61
Skaneateles	W. M. Beauchamp	1850-'67	Warrenton	Dr. W. M. Johnson	1857-'58
Sloansville	R. W. Potter	1868	Wilson	E. W. Adams	1868
Smithville	J. Everett Breed	1849-'52			
		1854-'58	**OHIO.**		
Somerville	Dr. P. H. Hough	1849-'51	Andrews	Dr. W. W. Sprall	1862-'61
South Edmeston	L. A. Beardsley	1849-'51	Athens	Prof. W. W. Mather	1849-'51
South Hartford	Granville M. Ingalsbee.	1863-'68	Austinburg	J. G. Dole and C. S. Griffing.	1862-'63
South Trenton	Capt. Storrs Burrows	1863-'69	Austinburg	David S. Alvord	1864
Spencertown	A. W. Morehouse	1855-'59	Austinburg	J. G. Dole	1864
Spencertown	Irving Magro	1858	Austinburg	E. D. Winchester	1864-'66

List of Smithsonian meteorological stations and observers—Continued.

Name of station.	Name of observer.	Years of observations.	Name of station.	Name of observer.	Years of observations.
OHIO—Continued.			OHIO—Continued.		
Avon	Rev. L. P. Ward	1858-'60	Hudson	Prof. C. A. Young and E. W. Childs.	1858-'59
Bellecomte	Rev. R. Shields and J. C. Smith.	1857-'59	Hudson	Prof. C. A. Young and A. C. Barrows.	1860-'61
Bellecentre	Rev. Robert Shields	1854 / 1860-'61	Hudson	Prof. C. A. Young, E. W. Haugrt, J. C. Elliot, W. Pettingill, H. R. Watterson.	1862
Bellefontaine	Joseph Shew	1855-'60			
Berea	Prof. G. M. Barber	1854			
Bethel	George W. Crane	1858-'60	Hudson	Prof. C. A. Young and J. C. Elliot.	1863
Bowling Green	W. M. Peck, M. D.	1852-'63			
Bowling Green	John Clarke	1858-'68			
Brookville	Rev. S. L. Hillier	1859-'61	Huron	Edmund W. West	1854
Cardington	Hubert A. Schauber	1863	Iberia	S. T. Boyd	1858
Centralia	Hubert A. Schauber	1854-'66	Jackson	Geo. L. Crookham	1849-'54
Clarviot	Ebenezer Hannaford	1855-'57	Jackson	G. L. Crookham and M. Gilmore.	1855
Cincinnati	John Lea	1849			
Cincinnati	P. W. Hartt	1854 / 1857-'58	Jackson	S. R. Wood	1855
			Jackson	M. Gilmore	1857-'58
Cincinnati	Geo. W. Harper	1855-'68	Jacksonborgh	J. B. Owsley, M. D.	1848
Cincinnati	A. A. Warder	1859-'63	Jefferson	James D. Herrick	1855-'58
Clockanal	R. C. & J. H. Phillips	1859-'64	Kerue	Dr. E. C. Bidwell	1849-'58
Cincinnati	Eli T. Tappan	1860-'69	Kenae	E. Spencer	1858-'54
Cincinnati	R. C. Phillips	1863-'68	Kenton	C. M. Smith, M. D.	1862-'63
Cleveland	Gustavus A. Hyde	1851			1864-'66
		1855-'61	Kelly's Island	Geo. C. Huntington	1858-'68
		1868	Kingston	Prof. Jno. Haywood	1865-'67
Cleveland	Edward Wade	1852	Lafayette	Samuel Knobie	1867
Cleveland	Edward Colborn	1858-'63	Laumster	Lewis M. Dayton	1857
Cleveland	U. S. Engineers	1860-'63	Lancaster	H. W. Jaeger	1858
Cleveland	G. A. Hyde and Mrs. Hyde.	1863-'67	Lancaster	W. C. Davis	1858
			Lancaster	J. W. Towne	1868
Cleveland	T. A. Smart	1866-'68	Lebanon	Joseph C. Hatfield	1849
College Hill	G. S. Ormsby	1854	Little Hocking	James Fraser	1862-'63
College Hill	Prof. E. S. Hoxworth	1853-'57	Little Mountain	L. J. Ferris	1857-'58
College Hill	Prof. J. H. Wilson	1858-'63	Madison	Rev. L. S. Atkins	1857-'60
College Hill	J. W. Hammitt	1858-'64	Madison	Mrs. Ardalia C. King	1858-'63
College Hill	L. R. Tuckerman	1853-'57	Mansfield	P. A. Renton	1851-'58
Collingwood	Henry Brewett	1856-'57	Marietta	D. P. Adams	1860-'63
Collingwood	Sarah E. Bartlett	1859	Marietta	Prof. J. W. Andrews	1848-'55
Cumberton	Thos. H. Johnson	1861-'62	Marion	H. A. True	1853-'64
Columbus	Thos. G. Wormley	1851-'59	Marion	T. Chase	1858
Croton	Mark Sperry	1860	Martin's Ferry	Charles R. Shreve	1867
Croton	Rev. E. Thompson and Mark Sperry.	1861	Medina	Rev. L. F. Ward	1852-'58
			Medina	Wm. P. Clarke	1858-'63
Croton	Rev. Elias Thompson	1862-'63	Middlebury	Michael Boerber	1849
Cuyahoga Falls	D. M. Henkie	1862-'63	Milnersville	Rev. D. Thompson	1862-'64
Dalkeith	F. G. Hill	1859-'63	Monroe county	Enoch D. Johnson	1859
Dayton	Cooper Female Seminary.	1856	Mount Auburn	Senior class Mt. Auburn Female Inst.	1868
Dayton	Jas. C. Fischer, M. D.	1856	Mt. Pleasant	David H. Tweedy	1859-'60
Dayton	Lewis Gruneweg	1854	Mount Tabor	William Lapham	1849
East Cleveland	Mrs. M. A. Pillsbury	1861-'63	Mount Union	Newton Anthony	1857-'60
East Fairfield	S. R. McMillan	1859-'67	Mount Vernon	P. A. Renton	1853-'55
East Rockport	Dr. J. P. Kirtland	1854	Mount Victory	W. C. Hampton	1859-'70
Eaton	Thomas J. Larsh	1863-'65	Newark	Lewis M. Dayton	1854-'55
Edinburg	Smith Ranford	1857-'58	Newark	Isaac Dille	1859-'63
Franklin	W. L. Schrock, M. D.	1855-'57	New Concord	Prof. R. G. Irvine	1849
Freedom	H. M. Davidson	1858-'60	New Lisbon	J. F. Benner	1857-'58
Freedom	H. M. Davidson and Wilson Davidson.	1861	New Westfield	A. E. Jerome	1862-'63
			North Bend	R. B. Warder	1844
Freedom	Wilson Davidson	1862	North Fairfield	O. Burrus	1857-'63
Gallipolis	G. W. Livesay	1854-'56	Northwood	Prof. J. R. W. Sloane	1858
Gallipolis	A. P. Rogers	1857-'58	Norton	W. D. Watkins	1849
		1864-'68	Norwalk	G. A. Hyde	1854
Garrettsville	Warren Pierce	1867-'63	Norwalk	Rev. Alfred Newton	1861-'68
Germantown	L. Gruneweg	1852-'56	Oberlin	Prof. Fairchild and Dascomb.	1849-'50
Germantown	J. S. Binkerd	1856-'57			
Granville	Prof. P. Carter	1849	Oberlin	Prof. J. N. Allen	1851-'52
Granville	Dr. S. N. Sanford	1849-'52			1857
Darwar	W. G. Fuller	1860-'61	Oberlin	Prof. J. H. Fairchild	1853-'56
Hillsborough	Rev. J. McD. Mathews	1851-'60	Oberlin	Frederick Allen	1859
		1863-'69	Perryburg	F. Holkerbeck	1854-'56
Hillsborough	C. C. James	1857			1859
Hillsborough	Dr. C. C. Simms	1863	Perryburg	F. and D. K. Hollenbeck.	1857
Hiram	S. L. Diller and S. M. Luther.	1855			
Hiram	Spencer L. Hillier	1856	Portsmouth	James H. Pott	1855-'59
Hiram	S. M. Luther	1856-'60	Portsmouth	D. R. Cotton, M. D.	1859-'61
Hocking Port	Dr. John Rhoades	1859-'68	Portsmouth	Lad. Engelbrecht	1862-'65
Homer	Thos. F. Withrow	1859	Republic	Stephen S. Dorsey	1851
			Richmond	Jacob N. Drustlem	1854-'55

List of Smithsonian meteorological stations and observers—Continued.

Name of station.	Name of observer.	Years of observations.	Name of station.	Name of observer.	Years of observations.
OHIO—Continued.			**PENN.—Continued.**		
Ripley	J. Ammen	1857-'61	Byberry	John W. Sherman	1860-'61
Ripley	Dr. G. Remback	1861-'67	Byberry	Isaac C. Martindale	1861-'67
Ripley	Mrs. M. M. Marsh	1847-'68	Canonsburg	Prof. J. R. Williams	1849
Russell Station	J. W. Gamble	1858-'63	Canonsburg	F. L. Stewart	1849
Sandusky	Thomas Niell	1854-'59	Canonsburg	Rev. Wm. Smith, D. D.	1855-'61
Savannah	Dr. John Ingrum	1854-'73			1861-'69
Saybrook	Rev. L. S. Arklas	1862-'63	Canonsburg	Charles Davis	1860
Saybrook	James B. Frazer	1864-'66	Canonsburg	Lyceum Jefferson Col.	1861-'63
Seville	Rev. L. F. Ward	1861-'62	Carlisle	Prof. S. F. Baird	1849
Sharonville	Wm. F. Bowen	1858-'60	Carlisle	Prof. W. C. Wilson	1855-'59
Sidney	Joseph Shaw	1857	Carlisle	W. H. Cook	1868
Smithfield	D. H. Tweedy	1860	Carpenter	F. L. McNeill	1862
Smithville	John H. Myers	1864-'68	Cerro	M. P. Stevens	1849-'54
Smithville	Wm. Hoover	1868	Chambersburg	Wm. Heyser, jr.	1858-'60
Steubenville	Berwell Marsh	1849-'63	Chromedale	Joseph Edwards	1854-'57
Steubenville	J. B. Doyle	1865-'69	Chromedale	Joseph Edwards and	1858
Springfield	Samuel O. Frey	1858-'61		John B. Smedley.	
Toledo	Sarah E. Brancil	1859	Clarksburg	Barus? McElroy	1852
Toledo	E. B. Hoffmesperger	1859	Connellsville	John Taylor	1849-'59
Toledo	J. B. Trembley, M. D.	1860-'68	Darby	John Jackson	1860-'69
Troy	Charles L. McClang	1859-'63	Dyberry	Theodore Day	1863-'69
Twinsburg	N. A. Chapman	1860	Easton	A. R. McCoy	1849
Unionville	Miss A. Cunningham	1854-'57	Easton	Prof. J. H. Coffin	1851
Unionville	Mrs. Ardelia C. King	1858	Easton	E. L. Dodder	1851
Urbana	Prof. M. G. Williams	1855-'69	Easton	Selden J. Coffin	1857-'58
West Bedford	H. D. McCarty	1856-'57	Easton	Selden J. Coffin and	1858-'60
Welshfield	B. F. Abell	1857-'66		G. N. Houghton.	
Wollington	Rev. L. F. Ward	1863	Easton	Geo. B. Houghton	1861
Westerville	Prof. Jno. Haywood	1858-'62	East Smithfield	James E. Tracy	1859
		1868	Erie	Benjamin Grant	1849
Westerville	Prof. H. A. Thompson	1863-'67	Ephrata	W. H. Speru	1863-'69
Western Star	A. S. Stiver	1861	Fallsington	Ebenezer Hance	1863-'69
West Union	Rev. Wm. Lumsden	1860-'61	Franklin	Rev. R. A. Tolman	1857-'62
Wooster	Eugene Pardee	1849	Freeport	Dr. A. Alter	1849
Wooster	Martin Winget	1864-'68	Freeport	Andrew Monleton	1848-'51
Williamsport	John R. Wilkinson	1867-'68	Freeport	A. D. Wier	1851
Windham	Samuel W. Treat	1857-'59	Freeport	John H. Baird	1860
Yankeetown	A. Jaycne	1854	Fleming	Samuel Brugger	1856-'67
Yellow Springs	W. A. Anthony	1863	Fountain Dale	S. C. Walker	1860
Zanesfield	John P. Lukins	1854	Germantown	R. Ebert	1859
Zanesville	L. M. Dayton	1858	Germantown	Thos. Meehan and J.	1863-'64
Zanesville	Adam Peters	1859		Meehan.	
Zanesville	J. O. F. Holston, M. D.	1853-'57	Germantown	Thomas Meehan	1858-'61
					1863-'69
OREGON.			Gettysburg	Prof. M. Jacobs	1848-'60
Albany	S. M. W. Hindman	1853-'59	Gettysburg	Rev. M. Jacobs and	1861
Auburn	R. B. Ironside	1862-'63		D. Eyster.	
Auburn	S. M. W. Hindman	1864-'65	Gettysburg	Rev. M. Jacobs and	1862-'63
Corvallis	A. D. Barnard	1868-'69		H. E. Jacobs.	
Fort Snyder	James A. Snyder	1858	Grampian Hills	Elisha Fenton	1854-'59
Fort Thompson	W. U. Wagner	1857-'58	Harrisburg	Dr. J. Heisely	1849-'50
Oregon City	Geo. A. Atkinson	1851-'58	Harrisburg	Wm. G. Hickok	1852-'54
Portland	Geo. H. Belman	1858-'59	Harrisburg	S. A. Martin	1860-'61
Salem	Thos. H. Crawford	1861	Harrisford	Dr. Paul Swift	1853-'59
Salem	P. L. Willis	1863-'65	Hollidaysburg	F. B. Lowrie	1853
			Honesdale	M. H. Cobb	1859
PENNSYLVANIA.			Horsham	Miss Anna Spencer	1864-'69
Abington	Rodman Simon	1864-'68	Huntingdon	Wm. Brewster, M. D.	1860
Altoona	W. H. Boyers	1858-'60	Ickesburg	Wm. E. Baker	1867-'69
Altoona	Thomas H. Savery	1863	Indiana	David Peeler	1849-'51
Andersville	R. Welser	1854	Indiana	Wm. D. Hildebrand	1858
Beaver	Rev. R. T. Taylor	1857-'68		and David Peeler.	
Bedford	Samuel Brown	1858-'59	Johnstown	David Peeler	1868
Bedford	Rev. H. Heckerman	1858-'61	Kingsleys	Francis Schreiber	1859
Bellefonte	J. L. Barrell	1858-'59	Lancaster	P. A. Muhlenburg, jr	1849
Hendersville	Franklin W. Cook	1859	Langaster	John Wise	1849-'51
Hendersville	T. E. Cook & Sons	1859-'60	Latrobe	Prof. Rudolph Muller	1860-'69
Berwick	John Eggert	1856-'61	Latrobe	W. B. Boyers	1861
		1863-'65	Lewisburg	Prof. C. S. James	1855-'60
Bethlehem	L. E. Harbaner	1849			1863-'69
Bethlehem	Nathan C. Tooker	1857	Lima	Messrs. Edwards and	1849-'59
Bethlehem	Prof. A. M. Mayer	1857-'69		Miller.	
Blairsville	W. B. Boyers	1861-'63	Lima	Joseph Edwards	1853
Blooming Grove	John Grathwohl	1863-'69	Lima	John H. Smedley	1859
Brookville	D. S. Deering	1854	Linden	James Barrell	1868-'69
Byberry	John Comly	1852-'54	Manchester	Corydon Marks	1849-'50
		1857-'58	Meadville	Prof. L. D. Williams	1849-'54
			Meadville	T. H. Thackran	1854-'58
			Media	Isaac M. Kerlin, M. D.	1862

List of Smithsonian meteorological stations and observers—Continued.

Name of station.	Name of observer.	Years of observations.	Name of station.	Name of observer.	Years of observations.
PENN.—Continued.			PENN.—Continued.		
Moorhead	R. L. Walker	1855	Tioga	E. T. Bratley	1853–56
Morrisville	Ebenezer Ilanco	1849–54	Towanda	S. J. Coffin, W. H.	1851
Morrisville	Mahlon Moore	1859		Dean, L. H. Kingsbury.	
Moss Grove	Francis Schreiner	1853–57			
Mount Joy	Mary E. Hoffer	1857–58	Troy Hill	Victor Bertha	1855–56
Mount Joy	Dr. Jacob R. Hoffer	1860–65	Uniontown	Freeman Lewis	1849
Murrysville	Thomas H. Stewart	1857–59	Valley Forge	C. P. Jones	1849
Murryville	F. L. Stewart	1867–69	Warrior's Mark	J. K. Lowrie	1854
Nazareth	E. T. Kluge	1651	Waynesboro'	Rev. L. J. Eyler	1853–54
Nazareth	E. T. Kluge and E. Kampmer.	1650	Wellsboro'	Henry W. Thorp	1849
			Westchester	Samuel Alsop	1858–59
Nazareth	H. A. Brickenstein	1855–57	Westchester	Prof. A.G. Clark and T. H. Aldrich.	1861–65
Nazareth	J. C. Harvey	1859–60			
Nazareth	O. T. Huebner	1861	Westchester	Dr. Geo. Martin	1868
Nazareth	O. T. Huebner and L. R. Bickeceler.	1869	Whitehall Station	Edward Kohler	1859–60
			Williamsport	H. C. Moyer	1864
Nazareth	L. E. Rickmeier	1863–66	Worthington	Samuel Scott	1856–62
New Castle	E. M. McConnell	1866–52	Youngsville	Dr. A. C. Dindget	1853–54
Norristown	Rev. J. O. Ralston	1851–53			
Northeast	John T. Milliken	1867	RHODE ISLAND.		
North Whitehall	Edward Kohler	1856–59			
Oil City	James A. Weeks	1863–64	East Greenwich	E. O. Arnold	1855–56
Oxford	Henry Duffield, M.D.	1863	Newport	Samuel Powel	1851
Paradise	Jacob Frantz	1854–58	Newport	Wm. H. Crandall	1863–62
Parkerville	Fenelon Darlington	1859–63	North Scituate	Henry C. Sheldon	1853–54
		1865	Portsmouth	Geo. Manchester	1851
Philadelphia	C. B. Navy-yard	1849	Providence	Prof. A. Caswell	1847–57
Philadelphia	Lt. Jos. Reed, U.S.N.	1849	Providence	H. C. Sheldon	1860–61
Philadelphia	Dr. Paul Swift	1849–52			
Philadelphia	J. F. Charlies	1848–52	SOUTH CAROLINA.		
Philadelphia	Prof. J. A. Kirkpatrick.	1856–60			
		1862–68	Aiken	H. W. Ravenel	1854–56
Philadelphia	U. S. Naval Hospital	1857–59	Aiken	Rev. J. H. Cornish	1857–61
Philadelphia	J. C. Martindale, M.D.	1860–61			1867–68
Philadelphia	P. Friel	1863	Anderson	E. S. Earle	1860
Philadelphia	Homer Eachers	1864	Barnwellville	Dr. Jno. P. Burrait	1851–53
Philadelphia	Penn'a mic Hospital	1864–65	Beaufort	Dr. M. M. Marsh and Mrs. Marsh.	1853–55
Philadelphia	J. M. Ellis	1867			
Pittsburg	Edward Fendrich	1849–51	Black Oak	Thos. P. Ravenel	1854–61
Pittsburg	Dr. H. Snyser	1849–50	Camden	J. A. Young. M.D.	1858–55
Pittsburg	W. W. Wilson	1852–56			1854–57
Pittsburg	Wm. Martin	1857	Camden	T. Carpenter	1851–54
Pittsburg	John Hastings and Wm. Martin.	1855	Charleston	Prof. L. R. Gibbes	1851
			Charleston	Dr. Jos. Johnson	1855–57
Pittsburg	Wm. Martin and Dr. Alex.M.Speer.	1859	Charleston	J. L. Dawson, M.D.	1857
			Charleston	Jos. Johnson, M. D. and J. L. Dawson, M. D., and G. R. Pelser, M. D.	1858–61
Pittsburg	Dr. Alex. M. Speer	1859–61			
Pittsburg	Prof. Rudolph Muller	1863			
Plymouth Meeting	Marcus H. Corson	1868			
Pocopson	Fenelon Darlington	1853–58	Columbia	Col. W. Wallace	1851
		1866–67	Columbia	F. H. Harleston	1856
Potterville	John Hughes	1854–55	Columbia	Prof. J. B. White	1856
Potterville	Dr. A. Heger	1855	Columbia	Capt. C. C. Tew	1858
Potterville	Rev. B. R. Smyser	1857	Columbia	E. J. Borton, M.D.	1858
Potterville	D. Washburn	1858	Columbia	Sup't Arsenal Acad'y.	1859
Randolph	Orrin T. Hobbs	1851–52	Edisto Island	F. N. Fuller	1855–57
		1854–56	Georgetown	Rev. Alex. Glennie	1859–61
Reading	John Hoyl Raser	1857–53	Gowdysville	Charles Petty	1863
		1866–56	Milton Hrod	Maj. J. W. Abott, U. S. Eng., Capt. G. R. Saint.	1861
Reading	Dr. J. B. Peale and Charles Hahn.	1858			
Scranton	Dr. A. P. Moybert	1859	Tilton Head	Maj. C. R. Suter, U. S. Engineers.	1863
Sewickleyville	John L. Travelli	1859–60			
Sewickleyville	J. L. Travelli and G. H. Tracy.	1861	Mount Pleasant	E. N. Fuller, M. D.	1857
			Orangeburg	Thos. A. Elliott	1849
Sewickleyville	George H. Tracy	1862	Orangeburg	Joseph T. Zealy	1849
Shannokin	P. Friel	1856–63	St. Johns	H. W. Ravenel	1848–21
Silver Spring	H.O. Brockart	1863–64	St. Johns	Thos. P. Ravenel	1850–61
Somerset	Rev. David J. Eyler	1859	Waccamaw	Rev. Alex. Glennie	1854–59
Somerset	Dr. F. Chorpenning	1859	Wilkesville	Chas. Petty	1855–57
Somerset	George Mowry	1857–59			
Somerville	J. Russell Dutton	1863–67	TENNESSEE.		
St. Mary's	Wm. A. Stokes	1849			
Sugar Grove	Lorin Blodget	1849–51	Austin	S. K. Jennings, M. D.	1857–61
Sugar Grove	W. O. Blodget	1850–54	Austin	P. R. Calhoun	1861
Summit Hill	M. Abbott	1858	Chattanooga	Dr. O. H. Blaker	1864
Summitville	Thos. Seabrook	1859	Clarksville	Prof. W. M. Stewart	1854–68
Susquehanna Depot	H. H. Atwater	1863	Dixon Springs	Thos. L. Sawyer	1852
Taneytown	John H. Baird	1856–60	Dover	B. F. Tavel	1859

List of Smithsonian meteorological stations and observers—Continued.

Name of station.	Name of observer.	Years of observations.	Name of station.	Name of observer.	Years of observations.
TENN.—Continued.			**TEXAS—Continued.**		
Elizabethton	Charles H. Lewis	1860	New Wied	T. C. Ervendberg	1853-'54
Fayetteville	Dr. W. W. McNulty	1849-'51	New Wied	J. L. Forke	1855-'58
Franklin	Jos. M. Parker, M.D.	1857	Pope's Expedition	James M. Reade	1855-'57
Friendship	Dr. Robert T. Carter	1854-'55	Port La Vaca	James Gardiner	1859
Greenville	N. R. & W. S. Doak	1856-'60	Roundtop	Bruno Sherman	1859-'61
Knoxville	O. W. Morris	1851-'52	San Patricio	J. O. Gaffney	1859-'60
Knoxville	Prof. Geo. Cooke	1853	Saterdale	Ernest Rapp	1859-'60
Knoxville	Prof. Geo. Cooke and L. Griswold.	1854	Springfield	T. A. Turner	1859
Knoxville	T. L. Griswold	1855-'56	Tehuat	Dr. B. L. D'Spain and J. M. Ewing.	1859-'60
Knoxville	Stephen C. Dodge	1860	Texana	William Colman	1859
La Grange	J. R. Blake	1859-'60	Turner's Point	James Rayal	1861
Lebanon	Prof. A. P. Stewart	1851-'54	Union Hill	Dr. Wm. H. Gent	1857-'61
Lebanon	Prof. B. C. Jillson	1854	Waco	Edward Merrill, M. D.	1857-'59
Lookout Mountain	Edward F. Williams	1866-'67			
Lookout Mountain	Rev. C. F. P. Bancroft	1867-'69	Washington	B. H. Rucker	1856-'60
Memphis	U. S. Navy Yard	1849,'53	Webberville	Prof. C. W. Yellowby	1858-'61
Memphis	R. Harris	1851-'52	Wheelock	F. Kellog	1859-'61
Memphis	W. J. Tuck, M. D.	1857-'5c	Woodboro	Dr. Jas. E. Moke	1859-'60
Memphis	Dr. Daniel F. Wright	1857			
Memphis	Drs. W. J. Tuck and R. W. Mitchell.	1859	**UTAH.**		
Memphis	R. W. Mitchell, M. D.	1860-'61	Gt. Salt Lake City	H. E. Phelps	1857
Memphis	Edward Goldsmith	1867-'68	Gt. Salt Lake City	M. K. Phelps and W. W. Phelps.	1858
Nashville	Prof. Jas. Hamilton	1848			
Nashville	Wm. Rothrock	1849	Gt. Salt Lake City	W. W. Phelps	1859-'61
Nashville	James Higgins	1854			
Nashville	Fred. H. French	1857-'58	Harrisburg	James Lewis	1859-'61
Pinson	J. W. Dodge & Son	1859-'61	Nebrville	Harrison Pearce	1859-'61
Trenton	Prof. Hamilton	1854	Rockville	Andrew L. Siler	1858
University Place, Franklin county.	Chas. H. Barney	1859-'61	St. George	Harrison Pearce	1858-'64
Walnut Grove	James R. Bean	1856-'57	St. George	H. Pearce and G. A. Burgon.	1865-'66
Winchester	S. W. Houghton	1859-'60	St. Mary's	Thomas Bullock	1865
			Vineland	Andrew L. Siler	1864
TEXAS.			Warship	Thomas Bullock	1858-'59
Aransas	Frederick Kalt	1860	Washington	Harrison Pearce	1859
Austin	Dr. Han'l K. Jennings	1852-'56			
Austin	J. W. Glenn	1854	**VERMONT.**		
Austin	Dr. R. K. Jennings and J. Van Nostrand	1857	Barnet	B. F. Eaton, M. D.	1856-'57
Austin	Swante Palm	1858-'61	Bradford	L. W. Liss	1856-'57
Austin	J. Van Nostrand	1856-'61 1867-'68	Brandon	D. Buckland	1852-'64
			Brandon	Harmon Buckland	1864-'67
Bastrop	J. D. Cunningham	1859	Brattleboro'	Charles C. Frost	1849-'51
Benham	Prof. Solomon Kiss	1859-'60	Brookfield	T. F. Pollard	1863
Boston	G. Pierce	1860-'61	Burlington	Prof. Zadok Thompson.	1849-'54
Boundary Survey	John H. Clark	1859			
Burkeville	Dr. N. P. West	1856-'61	Burlington	McK. Petty	1857-'64
Cedar Grove N's.	Hennell Stevens	1867-'68	Calais	James K. Tuby	1861-'64
Chappell Hill	W. H. Gant	1866-'67	Carleton	D. Underwood	1852-'54
Columbus	Dr. W. O. De Graffenried.	1859	Charlotte	M. E. Wing	1862
			Craftsbury	Chas. A. J. Marsh	1853-'54
Cross Roads	F. S. Wade	1859-'60	Craftsbury	James A. Paddock	1855-'67
Dallas	John M. Crockett	1859	East Bethel	Charles L. Paine	1863
Galveston	Drs. C. H. Wilkinson, H. A. McComly, and others.	1867	East Montpelier	B. J. Wheeler	1853
			Lunenburg	Hiram A. Cutting	1858-'69
			Middlebury	Prof. W. H. Parker	1849-'50
			Middlebury	Harmon A. Sheldon	1853-'69
Geological Survey	Geo. G. Shumard	1859	Montpelier	D. P. Thompson	1849,'51
Gilmer	J. M. Hearn	1859-'61 1867-'69	Montpelier	R. M. Marsh	1853
			North Craftsbury	Rev. Edward P. Wild	1857-'69
Ordiad	John C. Brightman	1857-'58	Norwich	A. Jackmap	1853-'59
Gonzales	Melvin H. Allis	1859-'61	Randolph	R. M. Menky	1849-'51
Greenville	Dr. R. De Jernett	1859-'60	Randolph	Charles L. Paine	1854-'69
Hebron	John C. Brightman	1856-'57	Rupert	Joseph Parker	1857-'60
Houston	Dr. A. M. Potter	1849-'53	Rutland	S. O. Mead	1862-'64
Houston	Miss E. Baxter	1867-'69	Saxes' Mills	J. C. Baker	1855
Huntsville	H. Yoakum	1849-'51	Shelburne	George Bliss	1855-'57
Huntsville	J. H. Browne	1858	Springfield	Rev. J. W. Chickering	1860-'63
Huntsville	T. Gibbs	1856-'60	St. Johnsbury	J. K. Colby and J. P. Fairbanks.	1853-'55
Kaufman	James T. Rayal	1859-'58			
Kanfman	James Brown	1859	St. Johnsbury	Franklin Fairbanks	1857-'61
Jefferson	W. T. Epperson	1859	West Fairlee	L. W. Bliss	1856
Larissa	F. L. Yoakum	1859-'60	Wilmington	Rev. John B. Perry.	1856-'67
Long Point	M. Rutherford	1857	Woodstock	Charles Marsh	1853-'59
New Braunfels	A. Forke and Otto Friedrich.	1857	Woodstock	Lester A. Miller	1867
New Braunfels	Otto Friedrich	1859-'60	Woodstock	H. Bodee and L. A. Miller.	1868

List of Smithsonian meteorological stations and observers—Continued.

Name of station.	Name of observer.	Years of observations.	Name of station.	Name of observer.	Years of observations.
VIRGINIA.			**WASHINGTON TERRITORY.**		
Anna	Rev. C. B. McKee	1858-'59	Fort Colville	Capt. Hague	1860-'61
Alexandria	Benj. Hallowell	1849	Fort Stellacoom	David Walker, M. D.	1853-'64
		1853-'59	Fort Vancouver	Dr. Barnes	1859
Ashland	Samuel Coarb	1874	Neeah Bay	James G. Swan	1862-'63
		1856-'57	Neeah Bay	Alexander Sampson	1867
Berryville	Miss E. Kownslar	1855-'57	Port Townsend	S. S. Dalkiop	1857-'68
Berryville	Dr. H. Kownslar	1858			
Bridgeton	C. R. Moore	1864	**WEST VIRGINIA.**		
Bridgewater	Jed. Hotchkiss	1852, '54			
Buffalo	Prof. G. R. Rossiter	1851-'54	Ashland	Charles L. Roffo	1855-'59
Buffalo	Samuel Coach	1855, '56	Barnum Springs	Robert H. Bliven	1857-'64
Buffalo	Wm. R. Boyers	1854-'59	Capon Bridge	J. J. Offutt, M. D.	1857
Cape Charles	Jeay G. Potts	1867-'68	Crackwhip	D. M. Ellis	1856-'57
Charleston	Jas. E. Kendall	1856-'57	Grafton	D. W. H. Sharp	1867-'68
Charlottesville	Chas. J. Meriwether	1849-'51	Hampshire county	S. J. Stumps	1868
Charlottesville	J. Ralls Abell	1859-'61	Harper's Ferry	L. J. Bell	1860
Christiansburg	Wm. C. Hagan	1851	Hunterville	Wm. Skeen	1851-'59
Cobham	Chas. J. Meriwether	1859	Kanawha	David L. Ruffner	1856-'57
Cobham Depot	Geo. C. Dickinson	1859-'61	Kanawha	James E. Kendall	1859
Crichton's Store	R. F. Astrop	1859-'61			1860-'61
Diamond Grove	R. F. Astrop	1849-'51	Kanawha Salines	W. C. Reynolds	1856-'59
Falmouth	Abraham Van Doren	1859-'61	Lewisburg	Dr. Wm. N. Patton	1851-'59
Fredericksburg	Chas. H. Roby	1859-'61	Lewisburg	Dr. Thos. Patton	1853-'57
Fredericksburg	B. R. Wellford	1849	Lewisburg	Thos. Patton and J.	1859
Fork Union	Silas H. Jones	1859-'61		W. Stalmaker	
Garysville	T. S. Beckwith, M. D.	1859	Lewisburg	J. W. Stalmaker	1859-'61
Garysville	Julian C. Raffa	1859	New Creek Station	Hendricks Clark	1856-'63
Gosport	United States Navy	1849	Point Pleasant	W. R. Boyers	1859
	Yard.		Romney	Marshall McDonald	1859
Hartwood	Abraham Van Doren	1858	Romney	W. H. McDowell	1866-'68
Hentleyville	J. C. Wills	1849	Sistersville	Enoch D. Johnson	1857-'59
Hewlett's	J. F. Adams	1867	Wardensville	D. H. Ellis	1855
Holliday's Cove	B. D. Sanders	1858			1859-'61
Johnstown	C. R. Moore	1849	Wellsburg	B. D. Sanders	1859-'60
Leesburg	N. F. D. Browne	1849	Weston	Benjamin Owen	1860
Leesburg	Samuel X. Jackson	1854	West Union	W. C. Quincy	1855-'59
Lexington	Wm. K. Park	1861	Wheeling	Geo. P. Lockwood	1859-'60
Lexington	W. D. Huffner	1867-'69	Wirt	Josiah W. Hoff	1856-'59
Lloyd	Geo. W. Upshaw	1859			
Longwood	Thos. J. Wickliss	1877	**WISCONSIN.**		
Lynchburg	A. Nettleton	1854			
Lynchburg	Chas. J. Meriwether	1858-'59	Appleton	Prof. R. Z. Mason	1856-'61
Madison	Dr. A. M. Grinnan	1851-'52	Appleton	John Hicks	1867
Meadow Dale	James Silvers	1857-'59	Appleton	Dr. M. J. E. Hurlburt	1867
Middlers	L. C. Brockenricin	1858	Appleton	Prof. J. C. Foye	1867-'68
Montcalm	Chas. J. Meriwether	1854	Ashland	Edwin Ellis	1849
Montrose	H. H. Fanniliroy	1856-'57	Antolan	James C. Brayton	1849-'52
Montrose	Edwd. E. Spence	1856-'59	Baraboo	Dr. B. F. Mills	1849-'52
Mossy Creek	Jed. Hotchkiss	1856-'58	Baraboo	M. C. Waite	1864-'69
Mount Solon	Jas. T. Clarke	1853-'56	Bay City	Edwin Ellis	1857-'58
		1857-'59	Bayfield	Harve. J. Nourse	1858-'59
Murtaphe	James Fraser	1857-'58	Bayfield	Andrew Tate	1867-'68
New England	James Fraser	1859-'61	Bellefontaine	Thomas Gay	1851-'54
Norfolk	United States Naval	1858	Beloit	Prof. S. P. Lathrop,	1849-'54
	Hospital.		Beloit	J. McQuigg and W.	1854
Portsmouth	N. R. Webster	1854-'60		Porter.	
Portsmouth	Naval Hospital	1860-'61	Beloit	Prof. Wm. Porter.	1855-'60
Powhatan Hill	Edward T. Taylor	1849-'59	Beloit	Prof. Henry S. Kenney	1861-'69
Prince Edward	Prof. Fra. J. Nullaner	1859-'52	Beloit	Henry D. Porter	1863-'67
Richmond	David Turner	1849-'54	Black River Falls	Emil Hauser	1859
Richmond	Chas. J. Meriwether	1859-'61	Brighton	George Matbewa	1862-'63
Richmond	John Appleyard	1860	Burlington	D. Matthews	1859
Rose Hill	Geo. W. Upshaw	1857-'59	Burlington	D. and G. Matthews	1859
Rougemont	Geo. C. Dickinson	1857-'59	Burlington	George Mathews	1861
Saltven	Julian C. Raffa	1856-'59	Caldwell Prairie	H. Armstrong	1860-'61
Salem	J. Carson Wells	1857-'59	Cascade Valley	M. Scoborn	1858
Smithfield	John B. Purdie	1856-'61	Cereseo	Miss M. E. Baker	1854-'55
Snowville	J. W. Shinkaker	1867-'68	Cereseo	M. H. Towers	1861-'62
Staunton	J. B. Imboden	1849	Delafield	Prof. A. W. Clark	1859-'60
Staunton	J. C. Covell	1858	Delafield	Chas. W. Kelly	1861-'63
Stribling Springs	Jed. Hotchkiss	1859	Delavan	Levon Eddy	1864-'67
Sorry	Benj. W. Jones	1857-'58	Edgerton	Henry J. Shinn	1867-'68
The Plains	John Pickett	1859-'60	Emerald Grove	Orrin Dinsmore	1849-'59
Winchester	J. W. Marvin	1852-'61	Embarras	J. Everett Head	1864
		1859-'61			1866-'68
Wytheville	W. D. Roedel	1861	Falls of St. Croix	M. T. W. Chandler	1857
Wytheville	Howard Shriver	1863-'68	Falls of St. Croix	Wm. M. Blanding	1858
Wytheville	Rev. Jas. A. Brown	1868	Galesville	Wm. Gale	1867-'68

List of Smithsonian meteorological stations and observers—Continued.

Name of station.	Name of observer.	Years of observations	Name of station.	Name of observer.	Years of observations
Wis.—Continued.			**Wis.—Continued.**		
Geneva	Wm. H. Whiting	1863–'69	New London	J. Everett Breed	1857–'58
Green Bay	Col. D. Underwood	1850	Norway	John E. Illmoe	1855–'57
Green Bay	Frederick Deckner	1864–'65	Osinah	Edwin Ellis, M. D.	1863–'66
Green Lake	C. F. Pomeroy	1851–'52	Otsego	L. H. Doyle	1859–'60
Hartford	Judge Hopewell Coon	1850–'69	Pardeeville	S. Armstrong	1859–'60
Hingham	John De Lyser	1867–'68	Platteville	Dr. J. L. Pickard	1851–'59
Hudson	O. F. Livingston	1854	Platteville	A. K. Johnson	1850–'51
Janesville	J. F. Willard	1853–'58	Platteville	G. Mosller	1863–'69
Janesville	Geo. J. Kellogg	1850	Plymouth	Spencer L. Hiller	1857
Janesville	Dr. Clark G. Pease	1800–'01	Prescott	Rev. Roswell Park	1858
Janesville	Daniel Strook	1862	Racine	W. J. Durham	1856–'59
Kenosha	Rev. John Gridley	1851–'59	Racine	Elland W. Phelps	1860–'61
		1857–'67	Racine	Prof. W. H. Ward	1863–'66
Kilbourn City	James H. Bell	1863–'64	Ripon	Prof. W. H. Ward	1863–'66
Lake Mills	Isaac Atwood	1859–'62	Rocky Run	W. W. Carns	1859–'63
Lebanon	J. C. Hicks	1861	Rural	H. H. Struthers	1860–'61
Lied	H. H. Struthers	1858	Southport	Rev. John Gridley	1849
Madison	Prof. S. H. Carpenter	1863	Sumoll	Edward S. Spencer	1851–'54
Madison	S. H. Carpenter and J. W. Sterling.	1854	Superior	Wm. B. Newton and L. Washington.	1853
Madison	A. Rehna, M. D.	1856–'59	Superior	L. and H. Washington, and G. Loring, jr.	1856
Madison	Prof. J. W. Sterling	1856–'59	Superior	Wm. Mann	1859–'63
		1863–'65	Superior	G. R. Stuuis and E. H. Ely.	1859–'63
Madison	J. Jennings	1870			
Madison	Prof. J. W. Sterling and S. P. Clarke.	1860	Waterford	S. Armstrong	1863
			Watertown	William Ayres	1854
Madison	Prof. J. W. Sterling and W. Fellows.	1861–'69	Waukesha	Prof. S. A. Dean	1855–'56
					1856–'60
Manitowoc	Jacob Lüps	1857–'68	Waukesha	Prof. S. A. Dean and L. C. Slye, M. D.	1857
Menasha	Col. D. Underwood	1857–'58			
Milwaukee	I. A. Lapham	1849–'53	Waupacca	J. Everett Breed	1858, '63
		1854	Trempacca	H. C. Mead	1863–'64
		1857–'68	Waupaca	C. D. Webster	1867
Milwaukee	Carl Winkler, M. D.	1851–'57	Waunau	W. A. Gordon, M. D	1859–'60
Milwaukee	F. C. Pomeroy	1855–'59	Weyauwega	Melsar Parker	1860–'61
Milwaukee	Prof. E. P. Larkin	1859–'61	Weyauwega	William Woods	1861–'64
Mosinee	J. S. Peshley	1859	Weyauwega	John C. Hicks	1864
Mount Morris	Wm. F. Horsford	1864	Weyauwega	* Dr. Jas. Matthews	1864
New Holstein	Ferdinand Huebes	1865	Whitewery	Edwin Ellis, M. D.	1859–'60
New Lisbon	John L. Danogan	1867–'68			

Alphabetical list of meteorological observers of the Smithsonian Institution, up to the end of the year 1868.

Name.	State.	Name.	State.
Abbe, Cleveland	Michigan.	Astrop, R. F.	Virginia.
Abbott, M.	Pennsylvania.	Atkins, Rev. L. S.	Ohio.
Abell, B. F.	Ohio.	Atkinson, George A.	Oregon.
Abell, J. Ralls	Virginia.	Atkinson, William A.	Iowa.
Abert, Major J. W.	North Carolina.	Atwater, H. H.	Pennsylvania.
Abert, Thayer	Florida.	Atwood, Isaac	Wisconsin.
Acadia College	Nova Scotia.	Aubier, Rev. J. M.	New York.
Adams, F. L.	Massachusetts.	Austin, W. W.	Indiana.
Adams, Prof. E. W.	North Carolina.	Avery, Charles A.	New York.
Adams, I. R.	New York.	Ayres, W.	Wisconsin.
Adams, J F.	Virginia.	Ayres, Dr. W. O.	California.
Adams, J. F.	Georgia.	*Babcock, A. J.	Illinois.
Adams, Jno. W.	Maine.	Babcock, Dr. D. F.	Minnesota.
Adams, W. H.	Illinois.	Babcock, E.	Illinois.
Adams, W P.	New York.	Babcock, E.	Iowa.
Agnew, J. C.	Missouri.	Bachelder, F. L.	Florida.
Agricultural College	Kansas.	Bachelder, J.	Massachusetts.
Alba, E. M.	New York.	Bacon, D. G.	Kansas.
Alcott, W. P.	Massachusetts.	Bacon, E. E.	Illinois.
Aldrich, T. H.	Pennsylvania.	Bacon, Frank M.	Michigan.
Aldrich, Verry	Illinois.	Bacon, William	Massachusetts.
Alexander, Arch	New Jersey.	Baer, Miss H. M.	Maryland.
Alexander, Captain R. E..	Bermuda.	Baer, Prof. W.	Maryland.
Alexander, J. M.	New York.	Bailey, James D.	Florida.
Alison, H. L.	Alabama.	Bailey, S. S.	Missouri.
Allan, Edwin	New Jersey.	Bailey, Thomas	Massachusetts.
Allen, Prof. E. A. H.	New York.	Baird, John H.	Pennsylvania.
Allen, Frederick	Ohio.	Baird, Prof. S. F.	Pennsylvania.
Allen, George D.	Florida.	Baker, Frank	Illinois.
Allen, Prof. O. N.	Ohio.	Baker, J. C.	Vt. & Canada.
Allen, James, jr.	Michigan.	Baker, Miss M. E.	Wisconsin.
Allen, J. P.	Nebraska.	Baker, N. T.	Illinois.
Allen, W. D.	New York.	Baker, William E.	Pennsylvania.
Adlin, Lucius C.	Massachusetts.	Baldwin, Dr. A. S.	Florida.
Atlis, Melvin H.	Texas.	Baldwin, Elmer	Illinois.
Allison, Jesse	Illinois.	Ball, Miss Ida E.	Iowa.
Allison, Thomas A.	North Carolina.	Ball, Dr. J. E.	Iowa.
Alsop, Samuel	Pennsylvania.	Ball, James H.	New York.
Alter, Dr. D.	Pennsylvania.	Ballon, N. E.	Illinois.
Alvord, D. S.	Ohio.	Bambach, Dr. G.	Ohio.
Ammen, J.	Ohio.	Bancroft, Rev. C. F. P.	Tennessee.
Anderson, C. F.	Minnesota.	Bandelier, Adolphus F., jr.	Illinois.
Anderson, H. H.	Indiana.	Bannister, H. M.	Alaska Ter.
Anderson, Dr. James	Georgia.	Barbage, Joshua C.	Kentucky.
Anderson, W, H.	Indiana.	Barber, Prof. G. M.	Ohio.
Andrews, G. P.	Michigan.	Barker, Thomas M.	Alabama.
Andrews, Prof. J. W.	Ohio.	Barlow, Dennis	Arkansas.
Andrews, Seth L.	Michigan.	Barnard, Alonzo	Minnesota.
Andrus, W. C.	Florida.	Barnard, A. D.	Oregon.
Angell, D. D.	Indiana.	Barnes, C.	Indiana.
Anthonioz, B. F.	Louisiana.	Barnes, Dr.	Washington Ter
Anthony, Newton	Ohio.	Barney, Charles R.	Tennessee.
Appleyard, John	Virginia.	Barratt, Rev. J. P.	South Carolina.
Arden, Thos. B.	New York.	Barrett, James	Pennsylvania.
Armstrong, M. K.	Dakota Ter.	Barrows, A. C.	Ohio.
Armstrong, S.	Wisconsin.	Barrows, G. B.	Maine.
Arnold, E. G.	Rhode Island.	Barrows, N.	Massachusetts.
Arnold, James B.	Bermudas.	Barrows, Storrs	New York.
Arnold, Mrs. J. T.	Georgia.	Bartlett, E. H.	New York.
Astron. Observatory, Williams College	Massachusetts.	Bartlett, Isaac	Indiana.
		Bartlett, Joshua	Maine.

Alphabetical list of meteorological observers, &c.—Continued.

Name.	State.	Name.	State.
*Barton, Dr. E. H.	S. Car. and La.	Bly, E. H.	Wisconsin.
Bassett, G. R	Illinois.	Board of Trade	Nova Scotia.
Batchelder, F. L.	Florida.	Boardman, G. A.	Florida.
• Baxter, Miss E.	Texas.	Boerner, C. G.	Indiana.
Beal, Dexter	Iowa.	Boettner, Gustave A.	Illinois.
Beal, W. W	Iowa.	Bogert, W. S.	Florida.
Beaman, Carlisle D.	Iowa.	Bond, Isaac	Maryland.
Bean, Dr. James D	Fla. and Tenn.	*Bond, W. C.	Massachusetts.
Bean, Prof. S. A	Wisconsin.	Bosworth, Prof. R. S.	Ohio.
Beans, Thomas J	New Jersey.	Boucher, W. K	California.
Beardsley, L. A	New York.	Bowen, Miss Anna M. J.	Kansas.
Beatty, O	Kentucky.	Bowen, J. S	Kansas.
Beauchamp, W. M.	New York.	Bowen, Wm. F	Ohio.
Beckwith, E. W	Mississippi.	Bowles, Dr. S. D	Missouri.
Beckwith, Dr. T. F	Virginia.	Bowlsby, G. W.	Michigan.
Beckwith, W	Kansas.	Bowman, E. H.	Illinois.
Beecher, Michael	Ohio.	Bowman, John	New York.
Behmer, Frederick	Missouri.	Boyd, S. T	Ohio.
Belcher, W. C.	California.	Boyers, W. R	Pennsylvania.
Belfield H. H	Iowa.	Boyle, C. R	Iowa.
Bell, Miss E. M. A	Illinois.	Brackett, G. E	Maine.
Bell, Jacob E	Maryland.	Brayton, Jas. C	Wisconsin.
Bell, James H	Wisconsin.	Brayton, Milton	Sombrero Isl'd.
Bell, J. J	Maine.	Breckenstein. L. C	Virginia.
Bell, Louis	New Hampshire.	Breed, J. Everett	N. Y. and Wis.
Bell, L. J	Virginia.	Breed, M. A	Illinois.
Bell, Lewis J	Maryland.	Brendel, Dr. E.	Illinois.
Bell, Hon. S. N	New Hampshire.	Brendel, Dr. F.	Illinois.
Benagh, George	Alabama.	Brewer, Rev. Fisk P.	North Carolina.
Benner, J. P	Ohio.	Brewer, F. A	Massachusetts.
Bennett, Henry	Ohio.	Brewster, Alfred	New Hampshire.
Bennett, Sarah E	Ohio.	Brewster, Wm., M. D.	Pennsylvania.
Benton, F. A	Ohio.	Brickenstein, Rev. H. H.	Illinois.
Berendt, G	Mexico.	Brinkerhoff, G. M	Illinois.
Berger, M. L	Massachusetts.	Briggs, E. L.	Iowa.
Berky, W. H	Kansas.	Brightman, J. C	Texas.
Berthoud, E. L	Ky..Kan.,&Col.	Brooks, Rev. Jabez	Minnesota.
Betts, Charles	Michigan.	Brooks, Hon. J	Massachusetts.
Bickford, Calvin	Maine.	Brooks, Wm	New Jersey.
Bidwell, Dr. E. C	Iowa and Ohio.	Brookes, Samuel	Illinois.
Binkerd, J. S	Ohio.	Brown, Branch	New Hampshire.
Birney, James O.	Michigan.	Brown, B. G	New Hampshire.
Bixby, A. H	N.H.,Ky.,&Ind.	Brown, E. E.	Maine.
Bixby, J. H	Massachusetts.	Brown, G. W.	Kansas.
Blackman, W. J. R.	Kansas.	Brown, H. H	Kansas.
Blackwell, Thomas	Canada.	Brown, James	Texas.
Blackwell, W. H	Arkansas.	Brown, Rev. J. J	New York.
Blake, H	Indiana.	Brown, J. W	Illinois.
Blake, J. R	Tennessee.	Brown, N. W	Massachusetts.
*Blaker, Dr. G. H	Tenn. and Mich.	Brown, Prof. P. P	Illinois.
Blakeslee, Rev. S. V	California.	Brown, P. P	Ind. Ter.
Blanchard, Orestes A	Illinois.	*Brown, Samuel	Pennsylvania.
Blanding, William M	Wisconsin.	Brown, W. H. G	New Hampshire.
Blewitt, Rev. W	Georgia.	Browne, J. H	Texas.
Bliss, George	Vermont.	Browne, N. F. D	Virginia.
Bliss, L. W	Vermont.	Brown, O. H	Kansas.
Bliven, Robert H	West Virginia.	Buchart, H. G	Pennsylvania.
Blodgett, Dr. A. C	Pennsylvania.	Brugger, S	Pennsylvania.
Blodget, Loris	Pennsylvania.	Bryant, A. F	Iowa.
Blodget, W. O	Pennsylvania.	Bryant, C. H	Illinois.
Bloodgood, S. De Witt	New York.	Buck, Rufus	Maine.

List of meteorological observers, &c.—Continued.

Name.	State.	Name.	State.
*Buckland, D..............	Vermont.	Chapman, J. P............	New York.
Buckner, Rev. H. F.......	Arkansas.	Chapman, N. A..........	Ohio.
Bulkeley, S. S...........	Wash. Ter.	Chappelsmith, J	Indiana.
Bullard, Ransom.........	Michigan.	Chase, Arthur...........	New Hampshire.
Buller, E. A.............	Maine.	Chase, Dr. Charles.......	New Hampshire.
Bullock, J. T	Indiana.	Chase. Cornelius	New York.
Bullock, Thomas.........	Utah.	Chase, C. Thornton......	New York.
Burgon, G. A	Utah.	Chase, Milton...........	Michigan.
Burras, O	Ohio.	Chase, T	Ohio.
Burrell, J. I	Pennsylvania.	Cheney, W	Minnesota.
Borris, Dr. Robert.......	Arkansas.	Chickering, J. W........	New York.
Burroughs, R...........	Indiana.	Chickering, Rev. J. W...	Vermont.
Burt, Henry M..........	Nebraska.	Child, Dr. A. L...../...	Nebraska.
Burton, John	Delaware.	Child, Miss J. E........	Nebraska.
Bush, Rev. Alva........	Iowa.	Childs, E. W...........	Ohio.
Bussing, D. S...........	New York.	Chorpenning, Dr. F	Pennsylvania.
Bussing, J. W	New York.	Christian, John	Missouri.
Butterfield, W. W.......	Indiana.	Clark, Prof. A. G.	Pennsylvania.
Butterfield, Mrs........	Indiana.	Clark, Prof. A. W	Mary'l'd & Wis.
Byers, Dr. Robert	Missouri.	Clark, Dr. D......	Michigan.
Byers, S. M.............	Minnesota.	Clark, Sereno...........	New York.
Byers, William N	Nebraska.	Clark, Hendricks.	West Virginia.
Byram, E. N............	New York.	Clark, Thomas.........	Minnesota.
Byrne, Arthur M........	Illinois.	Clark, W. P............	Ohio.
Calder, J. G	Bermudas.	Clarke, John...........	Ohio.
Caldwell, John H	Massachusetts.	Clarke, J. T............	Virginia.
Caldwell, John T........	Missouri.	Clarke, Lawrence, jr.	Hud. Bay Ter.
Caldwell, R. H..........	Kentucky.	Clarke, Mrs. Lawrence..	Hud. Bay Ter.
Calhoun, P. B...........	Tennessee.	Clarkson, Rev. D.......	Kansas.
Calloway, J. W..........	North Carolina.	Cleland, Rev. T. H......	Kent'y & Miss.
Camp, Benjamin F	Georgia.	Clement, John, jr........	New Jersey.
Camp, E. P.............	Kansas.	Clough, J. D............	Minnesota.
Campbell, J	Missouri.	Cockburn, S............	Honduras.
Campbell, Dr. W. M.....	Michigan.	Cobb, M. H............	Pennsylvania.
Canfield, Dr. C. A.......	California.	Cobb, M. H............	Connecticut.
Cantril, J. E............	Illinois.	Cobbs, Rev. R. A.......	Alabama.
Canndas, Ant...........	Guatemala.	Cobleigh, Prof. N. E.	Illinois.
Capen, E	Illinois.	Coffin, Prof. J. H.......	Pennsylvania.
Carey, Daniel...........	Illinois.	Coffin, Matthew........	Michigan.
Carll, Albert G	New York.	Coffin, Selden J	Pennsylvania.
*Carothers, A. G........	Bahamas.	Coffin, Prof. William	Illinois.
Carpenter, D...........	Iowa.	Cofrau, L. B...........	Maryland.
Carpenter, Prof. S. H.....	Wisconsin.	Colburn, Ed...........	Ohio.
Carpenter, S. O.........	New York.	Colby, Alfred	New Hampshire.
Carpenter, T...........	South Carolina.	Colby, E. P............	New Hampshire.
Carter, J. H	Louisiana.	Colby, J. K............	Vermont.
Carter, Rev. J. P........	Maryland.	Cole, Barker...........	New Jersey.
Carter, Prof. P.........	Ohio.	Cole, E. C.............	New Jersey.
Carter, Dr. Robert T.....	Tennessee.	Coleman, J. A..........	Alabama.
Carter, Thomas	North Carolina.	Collier, Alfred.........	Massachusetts.
Carver, Rev. E. W.......	Minnesota.	Collier, D. C...........	Colorado.
Carver, Dr. R. T.......	Tennessee.	Collier, Prof. George H...	Illinois.
Case, Dr. C. D	Kentucky.	Collin, Prof. Alonzo	Iowa.
Case, Jarvis	Connecticut.	Collins, Rev. Samuel.....	Indiana.
Caswell, Prof. A.........	Rhode Island.	Collins, Colonel W. O....	Idaho.
Caswell, Rev. R. C	Newfoundland.	Colman, William........	Texas.
Cavileer, Chas	Minnesota.	Comings, G. P..........	Missouri.
Chadbourne, Prof. P. A...	Mass. & Conn.	Comly, John...........	Pennsylvania.
Chamberlain, J	Iowa.	Conant, Marshall.......	Massachusetts.
Chandler, Charles Q......	Missouri.	Conkling, Thomas J.....	Missouri.
Chandler, Dr. George	Massachusetts.	Connolly, H...........	Hud. Bay Ter.
Chandler, M. T. W.......	Wisconsin.	Cook, E. B............	New York.

List of meteorological observers, &c.—Continued.

Name.	State.	Name.	State.
Cook, Ephr. H	New Jersey.	Darlington, F	Pennsylvania.
Cook, F. W	Pennsylvania.	Dascomb, Professor	Ohio.
Cook, Prof. George H	New Jersey.	Davidson, H. M	Ohio.
Cook, Thomas E., & Sons	Pennsylvania.	Davidson, W	Ohio.
Cook, W. H	Pennsylvania.	Davies, James F	Arkansas.
Cooke, Prof. George	Tennessee.	Davis, Charles	Pennsylvania.
Cooke, R. L	New Jersey.	*Davis, Rev. Dr. Emerson	Massachusetts.
Cooper Female Seminary	Ohio.	Davis, George L. C	Mississippi.
Cooper, Dr. George F.	Georgia.	Davis, R. J	Virginia.
Coorlies, J. F	Pennsylvania.	Davis, W. E	Ohio.
*Corey, Henry M	Florida.	Dawson, John L	South Carolina.
Cornette, Rev. A	Alabama.	Dawson, W	Indiana.
Cornish, Rev. J. H	South Carolina.	Day, Theodore	Pennsylvania.
Corse, John M	Iowa.	Day, William	New York.
Corson, Rev. Marcus H	Pennsylvania.	Dayton, E. A	New York.
Cotting, J. M	Georgia.	Dayton, Lewis M	Ohio.
Cotton, Dr. D. B	Ohio.	Dayton, James H	Indiana
Couch, E. D	New Jersey.	Deacon, John C	New Jersey.
Couch, Samuel	Virginia.	Dean, W. H	Pennsylvania.
Cones, Dr. E.	Arizona.	Dearing, D. S	Pennsylvania.
Coulter, B. F	Arkansas.	Deckner, Frederick, & Son	Georgia & Wis.
Covell, J. C	Virginia.	Drom, D	Indiana.
Coventry, W. B	Indiana.	Deering, D. S	Iowa & Penn'a.
Covert, A. B	New York.	De Graffenreid, Dr. W. G.	Texas.
Cowing, Philo	New York.	De Jarnett, Dr. R	Texas.
Cox, Judge Hopewell	Wisconsin.	Dr Lyser, J	Wisconsin.
Craigie, Dr. W	Canada West.	*Delaney, E. M. J	Newfoundland.
Crandall, William H	Rhode Island.	*Delancy, John, jr	Newfoundland.
Crandon, F	Illinois.	Denig, E. T	Nebraska.
Crane, George W	Ohio.	Denison, H. L	Kansas.
Crane, Heber	Michigan.	*Denning, W. H	New York.
Craven, Rev. B	North Carolina.	Dennis, W. C	Florida.
Craven, Thomas J	Delaware.	Deilen, J. N	Ohio.
Crawford, R	Delaware.	*Dewey, Prof. C	New York.
Crawford, W. A	Delaware.	Denhurst, Rev. E.	Me., Mass., & Ct.
Crawford, T. H	Oregon.	De Wyl, N	Missouri.
Cribbs, J. H	Mississippi.	D'Spain, Dr. B. L	Texas.
Crisp, John F	Indiana.	Dickinson, George C	Virginia.
Crisson, J. C	Alabama.	Dickinson, James P	Iowa.
Crocker, Allen	Kansas.	Dickson, Walter	Hud. Bay Ter.
Crockett, John M	Texas.	Dill, John B	New York.
Croft, Charles	California.	Dille, Israel	Ohio.
Crookham, George L	Ohio.	Dinsmore, H. L	New York.
Crosby, J. B	Michigan.	Dismore, O	Wisconsin.
Crozier, Dr. E. S	Indiana.	Doak, S. S	Tennessee.
Crowther, Benjamin	Mexico.	Doak, W. S	Tennessee.
Cumming, S. J	Alabama.	Dodd, Prof. C. M	Indiana.
Cunningham, Miss Ardelia	Ohio.	Dodder, E. L	Pennsylvania.
Cunningham, G. A	Massachusetts.	Dodge, J. W., & Son	Tennessee.
Cunningham, J. D	Texas.	Dodge, Stephen C	Tennessee.
Currier, Alfred O	Michigan.	Dodson, B. D	Missouri.
Curtis, W. W	Wisconsin.	Dole, J. G	Ohio.
Cutler, Prof. E	Connecticut.	Dorat, Dr. Charles	San Salvador.
Cutting, Ephr	California.	Dorsey, S. S	Ohio.
Cutting, Hiram A	Vermont.	Dorweiler, P	Iowa.
Dall, W. H	Alaska Territ'y.	Doten, H	Vermont.
Dalrymple, A. P	Maryland.	Doughty, William H	Georgia.
Dalton, O. D	Missouri.	Doyle, Joseph B	Ohio.
Dana, William D	Maine.	Doyle, L. H	Wisconsin.
Darby, Prof. John	Georgia & Ala.	Draper, Dr. Joseph	Massachusetts.
Darling, L. A	Massachusetts.		

* Deceased.

*List of meteorological observers, &c.—*Continued.

Name.	State.	Name.	State.
Drew, Dr. F. P..........	Kansas.	Farnsworth, P. J........	Iowa.
Drummond, Rev. J. H....	Kansas.	Faunileroy, H. H.......	Virginia.
Drysdale, Robert H.	North Carolina.	Featherston, G. W	Arkansas.
Dubois, H. G., Jr.......	Connecticut.	Follows, H. B	New York.
Dodley, T............	Illinois.	Fellows, W	Wisconsin.
Duffield, Dr. E	Missouri.	Felt, C. W., and others..	Massachusetts.
Duffield, Rev. Dr. George..	Michigan.	Felt, John..............	New York.
Duffield, Dr. H......	Pennsylvania.	Female College, Arkadelphia	Arkansas.
Duncan, Rev. Alexander..	Illinois.	Fenderick, I. E	Pennsylvania.
Dungan, John L........	Wisconsin.	Fendler, Aug	Massachusetts.
Dunkaw, Mrs. E. S.....	California.	Fendler, Aug..........	Missouri.
Dunkum, W. L........	California.	Ferguson, G. F	Florida.
Dunwoody, W. P.	Iowa.	Ferris, E. J...........	Ohio.
Durham, W. J........	Wisconsin.	Finfrock, J. H	Idaho.
Dutton, J. Russell.....	Pennsylvania.	Finley, A. J............	Iowa.
Eschern, Homer........	Pennsylvania.	Finley, H. S...........	Iowa.
Earle, E. S......	South Carolina.	Finley, P. F...........	Arkansas.
Earle, J. W......	New York.	Finley, R. W..........	Missouri.
Earle, Silas........	California.	Fischer, Dr. James C.....	Ohio.
Easter, Prof. J. D......	Georgia.	Fish, Edmund	Kansas.
Eaton, Dr. D. F....	Vermont.	Fish, Lucien...........	Kansas.
Ebert, S............	Pennsylvania.	Fisher, Dr. J. C	Ohio.
Eddy, Levens	Wisconsin.	Fisher, Dr. Galen, M	Florida.
Edgerly, L. G.........	Illinois.	Fiske, W. M. L.........	New York.
Edwards, Daniel........	New York.	Fitch, Dr. Joseph	Illinois.
Edwards & Miller.......	Pennsylvania.	Fitzgerald, Rev. F	North Carolina
Edwards, Joseph........	Pennsylvania.	Fleming, John..........	New Jersey.
Edwards, Rev. Dr. Tryon..	Connecticut.	Flett, Andrew	Hud. Bay Ter.
Eggert, John..........	Pennsylvania.	Flint, Prof. W	New York.
Eldridge, Rev. W. V....	Illinois.	Flippin, W. B	Arkansas.
Elletson, Job.........	New York.	Flores, Thomas W	Mississippi.
Ellis, D. H............	Massachusetts.	Foote, Elisha	New York.
Ellis, D. H............	Virginia.	Folsom, C. A..........	Mississippi.
Ellis, J. M........	Pennsylvania.	Force, N. P............	Missouri.
Ellis, Dr. Edwin......	Mich. & Wis.	Foxcy, J. C............	Iowa.
Ellis, Dr. W. T.........	Colorado & Kas.	Forke, A	Texas.
Elliott, Jonathan......	St. Domingo.	Forke, J. C............	Texas.
Elliott, Prof. J. Lloyd...	Mississippi.	Foster, J. H	Michigan.
Elliott, J. C............	Ohio.	Foster, Robert W........	Louisiana.
Elliott, T. A..........	South Carolina.	Foster, W. L..........	Alabama.
Ellsworth, J.........	Illinois.	Foye, Prof. J. C	Wisconsin.
Ellsworth, Lewis.......	Illinois.	Franiz, Jacob..........	Pennsylvania.
Ellsworth, M. S.........	Illinois.	Fraser, James	Ohio.
Elston, W. J.........	Indiana.	Fraser, James	Virginia.
Engelbrecht, Lud	Ohio.	Fraser, James B	Ohio.
Engelmann, Dr. George..	Missouri.	Freeman, F. N.........	New Hampshire.
Epperson, W. T........	Texas.	Freeman, H. C.........	Illinois.
Ervendberg, Dr. L. C....	Mexico.	Freese, G	Texas.
Ervendberg, T. C.	Texas.	French, Fred. H	Tennessee.
Evans, John..........	Nebraska.	French, I. S	New Hampshire.
Eveleth, S...........	Maine.	French, J. D..........	New Hampshire.
Eveleth, Samuel A......	Maine.	French, J. H..........	New York.
Everett, Franklin......	Michigan.	Frey, Samuel C	Ohio.
Ewing, J. M..........	Texas.	Friedrich, Otto.........	Texas.
Eyler, Rev. D. J.......	Pennsylvania.	Friel, P	Pennsylvania.
Eyster, D	Pennsylvania.	Fries, G. W	New York.
Fairall, H. H........	Iowa.	Fritts, J. S	New Jersey.
Fairchild, Prof. J. H	Ohio.	Frombes, Prof. Oliver S...	California.
Fairchild, John P	New York.	Frost, Prof. A	New Jersey.
Fairbanks, F	Vermont.	Frost, C. C	Vermont.
Fairbanks, J. P..........	Vermont.	Frost, Colonel E. C	New York.
Fallon, John...........	Massachusetts.		

List of meteorological observers, &c.—Continued.

Name.	State.	Name.	State.
Fry, Lieutenant Joseph...	Florida.	Grant, C. W............	Illinois.
Fuller, Dr. E. N	South Carolina.	Grant, Miss Ellen	Illinois.
Fuller, R. M............	New York.	Grant, J...............	Illinois.
Fuller, W. G	Ohio.	Grant, Dr. W. T........	Georgia.
Gaffney. J. O...........	Texas.	Grape, George S.........	Maryland.
Gaines, Rev. A. G.......	Maine.	Grathwohl, J	Pennsylvania.
Gale. W	Wisconsin.	Grave, Mary A..........	Minnesota.
Gamble. J. W	Ohio.	Green, A. R	Mississippi.
Gantt, Dr. W. H........	Texas.	Greene, Prof. Dascom....	New York.
Gardiner, James	Texas.	Gregory, S. O...........	New York.
Gardiner, J. H.........	New York.	Gridley, Rev. J	Wisconsin.
Gardiner, R. H.........	Maine.	Griest, John	Indiana.
*Gardiner, Hon. R. H....	Maine.	Griest, Miss M	Indiana.
Garland, J. G..........	Maine.	Griffing, Henry	Illinois.
Garland, Samuel G	Ilabamas.	Griffith, R. H	Michigan.
Garrison, O. E.........	Minnesota.	Grinnell, J	Kentucky.
Garvin, Dr. P. C	Florida.	Griswold, L	Tennessee.
Gary, Jacob S	New Jersey.	Griswold, T. L.........	Tennessee.
Gay, Thomas	Wisconsin.	Groesbeck, Mrs. E. W ...	Kansas.
Gay, V. P.............	Illinois.	Groff, T. Louis	Illinois.
Gibbes, Prof. L. R......	South Carolina.	Groneweg, L	Ohio.
Gibbon, Lardner	Florida.	Guest, W. E..........	New York.
Gibbons, H............	California.	Gunn, Donald	Hud. Bay Ter.
Gibbons. J. S	New York.	Gupthll, G. W........	Maine.
Gibbs, T	Texas.	Haas, Henry	New York.
Gibson, H.............	Nebraska.	Habersham, S. E......	Georgia.
Gibson, R. T...........	Georgia.	Hachez, Ferd	Wisconsin.
Giddings, Rev. George D..	Illinois.	Hagan, W. F.........	Virginia.
Gidley, I. M	Iowa.	Hagenbick, J. M......	Iowa.
Gifford, D. R	Louisiana.	Hague, Captain	Washington Ter
Gifford, R. R	Massachusetts.	Hahn, Charles	Pennsylvania.
Giles, F. W	Kansas.	Haines, John	Indiana.
Gill, Joseph H	Illinois.	Haines, William	Georgia.
Gillingham, W.........	Maryland.	Hall, Dr. A..........	Canada.
Gilliss, Charles J......	Massachusetts.	Hall, Joel	Illinois.
Gilman, Stephen	Maine.	Hall, Joseph H	New Hampshire
Gilman, W. H	Kansas.	Hall, S. W	Maine.
Gilman, W. S. jr	New York.	Hallowell, Benjamin ...	Virginia.
Gilmore, M	Ohio.	Hamaker, Marion F	Missouri.
Gilmour, A. H. J	Canada.	Hamilton, Prof. J.....	Tennessee.
Glasco, J. M	Texas.	Hamilton, Captain W....	Ilabamas.
Glenn, J. M	Texas.	Hamilton, Rev. W	Nebraska.
Glennie, Rev. Alex	South Carolina.	Hamilton, W. H	North Carolina.
Glover, Ell S	Georgia.	Hammitt, J. W	Ohio.
Goff, Mrs. M. A	Michigan.	Hampton, W. C.......	Ohio.
Gold, T. S.............	Connecticut.	Hanau, B. P.........	Missouri.
Goldsmith, E	Tennessee.	Hance, Ebenezer	Pennsylvania.
Goodman, W. R........	Maryland.	Hanna, G. B	Massachusetts.
Goodnow, I. T.........	Kansas.	Hanniford, E........	Ohio.
Gordon, Robert	California.	Hanscom, R. F	New Hampshire
Gordon, Dr. W. A.......	Wisconsin.	Hanshew, Henry E	Maryland.
Gorton, J. S	New York.	Harding, Colonel Horace.	Alabama.
Goss, B. F............	Kansas.	Hardy, H. F	Arkansas.
Goss, George C	Iowa.	Hardy, Dr. J. F. E....	North Carolina.
Goss, W. K	Iowa.	Harkuess, W........	New York.
Gould, Lewis A.........	California.	Harper, G. W	Ohio.
Gould, M..............	Maine.	Harper, Prof. L......	New Jersey.
Graham, J.............	New York.	Harper, Prof. L......	Mississippi.
Graham, Paul	Arkansas.	Harris, A. J.........	Alabama.
Grant, Benjamin	Pennsylvania.	Harris, Dr. J. O.....	Illinois.
Grant, Charles	Georgia.	Harris, H...........	Tennessee.

List of meteorological observers, &c.—Continued.

Name.	State.	Name.	State.
Harrison, D. F...........	Connecticut.	Hollenbeck, F...........	Ohio.
Hart, J. H..............	New York.	Holley, Benj. F.........	Alabama.
Hart, C. F.............	Nova Scotia.	Hollingworth, G. W.....	Kansas.
Harvey, J. C..........	Pennsylvania.	Hollister, Dr. J........	Michigan.
Haskin, W. L..........	New York.	Holmes, D. J..........	Massachusetts.
Hastings, J...........	Pennsylvania.	Holmes, E. S.........	New York.
Haswell, Rev. J. R.....	New York.	Holmes, Dr. E. S......	Michigan.
Hatch, A. L...........	Mississippi.	Holmes, J. C.........	Michigan.
Hatch, Dr. F. W.......	California.	Holmes, Thomas......	Indiana.
Hatch, H..............	Mississippi.	Holston, Dr. J. G. F...	Ohio.
Hatch, J.............	New Hampshire.	Holt, Rev. Jos. W.....	Minnesota.
Hatfield, Jos. C......	Ohio.	Holt, William.........	Illinois.
Hanner, Ewil.........	Illinois & Wis.	Hoover, W...........	Ohio.
Haven, S. F..........	Massachusetts.	Hopkins, Prof. A.....	Massachusetts.
Hawks, Dr. J. M......	Florida.	Hopkins, Rev. Robert ...	Minnesota.
Hayes, Dr. W. W.....	California.	Hopkins, Prof. W. F...	Maryland.
Hayne, J. B..........	Bahamas.	Horn, Miss Clotilde......	Kansas.
Haywood, Prof. J	Ohio.	Horn, Dr. H. D.......	Kansas.
Heaston, D. J........	Missouri.	Horner, W. H........	Missouri.
Heckorman, Rev. H...	Pennsylvania.	Horr, Asa...........	Iowa.
*Hedges, Urban D	Delaware.	Horr, Ed. W..........	Iowa.
Hegar, Dr. A.........	Pennsylvania.	Horsford, William F....	Wisconsin.
Helmstreet, J. W.....	N. Y. and Minn.	Hosmer, A...........	New York.
Heisely, Dr. J........	Pennsylvania.	Hotchkiss, Jed.......	Virginia.
Helm, Thos. B........	Indiana.	Hough, Dr. F. B......	New York.
Henderson, W.........	Alabama.	Houghton, G. S.......	Pennsylvania.
Herrick, F. C.........	Kentucky.	Houghton, S. W......	Tennessee.
Herrick, J. D........	Ohio.	House, J. C..........	New York.
Hester, Lieut. J. W....	Florida.	House, J. Caroll.....	New York.
Heubner, O. T........	Pennsylvania.	Howard, J. 8........	Arkansas.
Heysor, W., jr.......	Pennsylvania.	Howe, Edwin........	California.
Hibbard, A. A........	N. Y. and Minn.	Howell, R...........	New York.
Hickok, W. O........	Pennsylvania.	Hoyt, Peter L........	New Hampshire.
Hicks, John.........	Wisconsin.	Hudson, Dr. A. T.....	Iowa.
Hicks, J. C..........	Wisconsin.	Hoehener, L. R......	Pennsylvania.
Hicks, Dr. W. R......	North Carolina.	Huestis, Prof. A. C...	Indiana.
Hieto, J. A..........	Mexico.	Huffaker, S, S.......	Missouri.
Higgins, Prof. D. F...	Nova Scotia.	Hughes, Jno.........	Pennsylvania.
Higgins, James......	Tennessee.	Hull, A. B..........	Connecticut.
*Hildreth, S. P......	Ohio.	Hunt, Ashley D......	Alabama.
Hill, F. G...........	Ohio.	Hunt, Rev. D........	Connecticut.
Hill, G. D...........	Dakota.	Hunt, George M......	New York.
Hill, L. J...........	Nebraska.	Hunt, Dr. S. B.......	New York.
Hillier, Rev. Spencer L ...	N.Y.,Ohio,Wis., and Minn.	Huntingdon, George C...	Ohio.
Himoe, John E.......	Wisconsin.	Hurlburt, Dr. M. J. E...	Wisconsin.
Himoe, Dr. S. O......	Kansas.	Hurin, Rev. W.......	New Hampshire.
Hindman, S. M. N.....	Oregon.	Hurtt, F. W.........	Ohio.
Hincox, G. D........	Illinois.	Huse, Frederick J....	Illinois.
Hoadley, C. H.......	Connecticut.	Huston, T. A........	Alabama.
Hobart, Ed. F.......	Iowa.	Hyde, G. A..........	Mass. and Ohio.
Hobbs, Chas. M......	Indiana.	Hyde, Mrs. G. A.....	Ohio.
Hobbs, Miss M. A.....	Indiana	Hyde, Stephen......	New York.
Hobbs, Orrin T.......	Pennsylvania.	Hyde, Rev. T. C. P....	New York.
Hobbs, William H.....	Indiana.	Imboden, J. D.......	Virginia.
Hoff, Dr. Alex. H....	Alaska Ter.	Ingalls, S. Marshall	New York.
Hoff, J. W...........	Virginia.	Ingalsbee, Granville M...	New York.
Hoffer, Dr. J. R......	Pennsylvania.	Ingersoll, J. D......	New York.
Hoffor, Mary E......	Pennsylvania.	Ingraham & Hyland	New York.
Hokcomb, A..........	Massachusetts.	Ingram, Dr. J.......	Ohio and N. J.
Hollenbeck, D. K.....	Ohio.	Ironside, N. D.......	Oregon.
		Irvine, Professor S. G ...	Ohio.

List of meteorological observers, &c.—Continued.

Name.	State.	Name.	State.
Ives, Edward B.	Florida.	*Kennicott, R.	Alaska & Hud.
Ives, William	New York.		Bay Territory.
Jackman, A	Vermont.	Kent, Benjamin	Massachusetts.
Jackson, Dr. A. W	Louisiana.	Kerlin, Dr. I. N	Pennsylvania.
Jackson, John	Pennsylvania.	Kerr, Prof. W. C	North Carolina.
Jackson, Samuel X	Virginia.	Kersey, Dr. V	Indiana.
Jacobs, H. E	Pennsylvania.	Kibbe, T. R	California.
Jacobs, Prof. M	Pennsylvania.	Kilgore, W	Minnesota.
Jaeger, H. W	Ohio.	Kilpatrick, Dr. A. R	Louisiana.
James, Miss Anna	Illinois.	King, Mrs. Adelia C	Ohio.
James, Prof. C. S	Pennsylvania.	King's College	Nova Scotia.
*James, Dr. John	Illinois.	Klugsbury, J. H	Pennsylvania.
Janes, C. C	Ohio.	Kirby, D. J	Massachusetts.
Jaque, A	Ohio.	Kirkpatrick, Prof. J. A	Pennsylvania.
Jonkins, John F	New York.	Kirtland, Dr. J. P	Ohio.
Jenkins, J. L	Illinois.	Kizer, W. B	Missouri.
Jennings, J	Wisconsin.	Kings, E. T	Pennsylvania.
Jennings, Dr. Samuel K	Ala., Texas, and	*Kings, Dr. J. P	New Granada.
	Tenn.	Knaner, J	Indiana.
Jilson, R. C	Tennessee.	Knox, James C	New Hampshire.
Johns, Dr. Montgomery	Maryland.	Knoud, Rev. James	Missouri.
Johnson, A. K	Wisconsin.	Kobler, Edward	Pennsylvania.
Johnson, Enoch D	Virginia & Ohio.	Kohlenkl, Theo.	Montana.
Johnson, E. W	New York.	Konlug, Rev. P. W	Missouri.
Johnson, Joseph	South Carolina.	Kownslar, Miss E	Virginia.
Johnson, Norton	Illinois.	Kownslar, Dr. R	Virginia.
Johnson, R. C	Nevada.	Kridelhaugh, S. H	Iowa.
Johnson, Rev. Samuel	New York.	Kron, F. J	North Carolina.
Johnson, Thomas H	Ohio.	Kummer, E	Pennsylvania.
Johnson, Warren	Maine.	Kunster, H	Illinois.
Johnson, Dr. W. M	North Carolina.	Laighton, Thomas B	New Hampshire.
Johnson, W. W	Montana.	Landon, Anna S	New York.
Johnston, Prof. John	Connecticut.	Landon, Stephen	New York.
Johnston, William M	Mississippi.	Lane, Prof. C. W	Georgia.
Jones, Benjamin W	Virginia.	Lane, John	Michigan.
Jones, C. P	Pennsylvania.	Lane, Moses	New York.
Jones, Prof. Gardner	Indiana.	Langdon, Laurens A	New York.
Jones, Josiah	Maryland.	Langee, Dr. Ignatius	Iowa.
Jones, Silas B	Virginia.	Langgutb, J. G., jr	Illinois.
Jones, W. Martin	Canada.	Langworthy, A. Darwin	Illinois.
Jorganson, C. N	Iowa.	Lapham, Darius	Ohio.
Jourdan, Prof. C. H	Maryland.	Lapham, I. A	Wisconsin.
Joy, Prof. Charles A	New York.	Larkin, Prof. E. P	Wisconsin.
Julien, Alexis A	Antilles.	Larrabee, W. H	Indiana.
Kalor, Frederick	Texas.	Larsb, T. J	Ohio.
Kapp, Ernest	Texas.	Laselle, C. B	Indiana.
Kaucher, William	Missouri.	Laszlo, Charles	Mexico.
Kedzie, Prof. R. C	Michigan.	Lathrop, Prof. S. P	Wisconsin.
Keenan, T. J. R	Mississippi.	Latimer, George	Porto Rico.
Keep, Rev. W. W	Florida.	Lattimore, Prof. S. A	New York.
Kellett, Thomas A	Minnesota.	Lawson, G. W	Dakota.
Kelley, O. H	Minnesota.	Lea, John	Ohio.
Kellogg, F	Texas.	Lear, O. H. P	Missouri.
Kellogg, G. J	Wisconsin.	Learned, Dwight W	Connecticut.
Kollum, A. O	Wisconsin.	Leavenworth, D. C	Connecticut.
Kelly, Charles W	Minnesota.	Lee, Charles A	New York.
Kelsey, Prof. H. S	Wisconsin.	*Lefferts, John	New York.
Kelsey, Kathela	New York.	Loffman, L	New York.
Kemper, G. W. H	Indiana.	Lefroy, Capt. J. H	Canada.
Kendall, James	Virginia.	*Lehman, Dr. H. M	Montana.
Kendall, John F	New York.	Leonard, Rev. L. W	New Hampshire.

List of meteorological observers, &c.—Continued.

Name.	State.	Name.	State.
Leonard, W. P............	Iowa.	McKee, Rev. C. B......	Dist. Col. & Va.
Levinga, Chas..........	Illinois.	McKenzie, J. M........	Iowa & Neb.
Lewis, Charles H........	Tennessee.	McLain, W. D..........	Colorado.
Lewis, James	Utah.	McMasters, Professor S. Y.	Ill's and Ken.
Lewis, James...........	New York.	McMillan, S. D........	Ohio.
Lingow, B. G...........	Missouri.	McMullen, J. F........	Minnesota.
Linus, M...............	New York.	McMore, P. A..........	New York.
Lippincott, James S.....	New Jersey.	McNelty, Dr. W. W.....	Tennessee.
Lippincott, Joseph W....	New Jersey.	McNett, E. L..........	Pennsylvania.
Little, Frank..........	Michigan.	McQuigg, J............	Wisconsin.
Little, J. Thomas......	Illinois.	McWilliams, Dr. Alex...	Maryland.
Livesay, G. W..........	Ohio.	Macfarlane, R.........	Ind. Bay Ter.
Livinga, B. C.........	Minnesota.	Macgregor, C. J.......	Canada.
Livingston, O. F......	Wisconsin.	Mack, A. W............	Massachusetts.
Livingston, Prof. Wm...	Illinois.	Mack, E. T............	New Jersey.
Lohdell, Mrs. M. J.....	New York.	Mack, Rev. E. T.......	New York.
Locke, Samuel..........	Minnesota.	Mack, Robert C........	New Hampshire
Lockhart, James........	Ind. Bay Ter.	Mackenzie, J..........	Ind. Bay Ter.
*Lockwood, G. P.......	Virginia.	Mackie, M.............	New York.
Logan, Dr. T. M........	California.	Magee, I..............	New York.
Loince, Dr. J. P.......	New York.	Magnetic Observatory ...	Canada.
Lord, W. G.............	Maine.	Maguire, Professor J. F..	Maryland.
Loring, C..............	Wisconsin.	Malcolm, W. S.........	New York.
Longbridge, Dr. J. H...	Indiana.	Malden, J. J..........	Michigan.
Love, Mrs. James.......	Iowa.	Mallinckrodt, C.......	Missouri.
Love, Louisa P.........	Iowa.	Manchester, Geo	Rhode Island.
Lowndes, B. O..........	Maryland.	Mankard, Mrs. M. J....	Louisiana.
Lowrie, J. R...........	Pennsylvania.	Manley, R. M..........	Vermont.
Lups, Jacob............	Wisconsin.	Mann, Wm..............	Wisconsin.
Lukins, John F.........	Ohio.	Mapes, H. H...........	Michigan.
Loll, James S..........	Mississippi.	Mar-y, Professor O ...	Illinois.
Lumpkin, W. M..........	Missouri.	Marks, Corydon........	Pennsylvania.
Lumsden, Rev. William...	Ohio.	Marlow, Colonel W. B...	Jamaica.
*Lunemann, John II.....	Missouri & Ky.	Marsh, Chas...........	Vermont.
Lupton, N. T...........	Alabama.	Marsh, C. A. J........	Vermont.
Luther, S. M...........	Ohio.	Marsh, M. M...........	Vermont.
Luttrell, James........	Colorado.	Marsh, Mrs. M. M......	Ohio.
Lynde, C. J............	Wisconsin.	Marsh, O. J...........	Illinois.
Lyons, Curtis J........	Massachusetts.	Marsh, Roswell	Ohio.
McAfee, J. R..........	Georgia.	Marshall, Gregory.....	Iowa.
McIlveth, Miss Sue.....	Arkansas.	Martin, Dr. Alex......	Indiana.
McCarly, H. D.........	Kansas & Ohio.	Martin, Dr. Geo.......	Pennsylvania.
*McCary, Robert.......	Mississippi.	Martin, Dr. G. Alex...	Arkansas.
McCary, William........	Mississippi.	Martin, K. A..........	Pennsylvania.
McClung, C. L..........	Ohio.	Martin, R. A..........	Delaware.
McComly, Dr. H. A.....	Texas.	Martin, Dr. Samuel D...	Kentucky.
McConnell, E. M........	Pennsylvania.	Martin, W.............	Pennsylvania.
McConnell, Townsend ...	Iowa.	Martindale, Dr. Jos. C..	Pennsylvania.
McCormick, J. O........	Maryland.	Martindale, Isaac C....	Pennsylvania.
McCormick, William A...	Kansas.	Marvin, J. W..........	Virginia.
McCoy, A. R...........	Pennsylvania.	Mason, Edgar E........	Nebraska.
McCoy, Dr. F	Iowa & Indiana.	Mason, Professor R. Z...	Wisconsin.
McCoy, Miss Lizzie	Iowa & Indiana.	Mather, Professor W. W..	Ohio.
McCready, Daniel.......	Iowa.	Mathews, D............	Wisconsin.
McDonald, M...........	Virginia.	Mathews, Geo..........	Wisconsin.
McDowell, Rev. A.......	North Carolina.	*Matthews, Dr. J......	Wisconsin.
McDowell, W. H........	West Virginia.	Mathews, Jos. McD	Ky. and Ohio.
McDowell, W. W........	North Carolina.	Mathis, H. C..........	Kentucky.
McElrath, J. J.........	Arkansas.	Matthews, H. W........	New York.
McElroy, Barnet	Pennsylvania.	*Matthews, Prof. M. M..	New York.
McGee, J..............	Massachusetts.	Mattison, Andrew	Kentucky.

* Deceased.

List of meteorological observers, &c.—Continued.

Name.	State.	Name.	State.
Manld, Dr. D. W	Delaware.	Moore, Jos	Indiana.
Mauncey, Rev. S. W	Minnesota.	Moore, Mahlon	Pennsylvania.
Mauran, Dr. P. B	Florida.	Moore, Dr. S. P	Louisiana.
Maurice, C. F	New York.	Moore, Dr. W	Missouri.
Maxey, W. F	Missouri	Mornhouse, A. W	New York.
Mayer, Professor Alfred M.	Md. & Pa.	Morelle, Professor D	North Carolina.
*Mayhew, Royal	Indiana.	Morris, Miss E	New York.
*Meacham H. G	Illinois.	Morris, Professor O. W.	New York.
Meacham, S	Illinois.	Morris, O. W	Tennessee.
*Mead, Allen	Iowa.	*Morrison, W. H	Illinois.
Mead, Chauncey	Iowa.	Morse, Geo. M	Massachusetts.
Mead, H. C	Wisconsin.	Morse, J. P	New York.
Mead, Dr. S. B	Illinois.	Mosca, J	Nicaragua.
*Mead, Stephen O	Vt. and N. H.	Moss, G. B	Illinois.
*Mead, Dr. Thompson	Illinois.	Moulton, J. P	Maine.
Meehan, J	Pennsylvania.	Moulton, M. M	Iowa.
Meehan, Thos	Pennsylvania.	Moultrie, J. L	Alabama.
Mecker, Ralph E	Illinois.	Mowry, Geo	Pennsylvania.
Memfield, Geo. C	Indiana.	Moyer, H. C	Pennsylvania.
Meriwether, C	Virginia.	Muhlenborg, F. A., jr	Pennsylvania.
Meriwether, C. J	Virginia.	Muir, Wm	Missouri.
Meriwether, R. T	Alabama.	Muller, Professor R	Pennsylvania.
Merriam, A. M	Massachusetts.	Mulligan, A	New York.
Merriam, A. M	Colorado.	Mulvey, Oliver	Indiana.
Merriam, C. C	New York.	Munfield, Geo. C	Indiana.
Merriam, G. F	Kansas.	Mungur, L. F	New York.
Merriam, Sidney A	Massachusetts.	Murch, E. M	Kentucky.
Merrick, Thos. B	New Jersey.	Murdoch, G	New Brunswick.
Merrill, Dr. E	La. and Texas.	Musery, R. D	Massachusetts.
Merrill, Rev. S. H	Maine.	Myers, J. H	Missouri.
Merritt, John C	New York.	Nash, Rev. J. A	Iowa.
Merwin, Mrs. E. H	Illinois.	Nason, Rev. Elias	N. H. and Mass.
Messman, Dr. J	Iowa.	Naturaliste Canadien	Canada.
Metcalf, H	New York.	Naval Hospital	California.
Metcalf, Dr. John G	Massachusetts.	Naval Hospital	New York.
Meybert, Dr. A. P	Pennsylvania.	Naval Hospital	Massachusetts.
Miles, Thos. H	Kentucky.	Naval Hospital	Pennsylvania.
Milnary Post	Newfoundland.	Naval Station	New York.
Millard, A. J	Iowa.	Navy Yard	Florida.
Millard, Joseph D	Michigan.	Navy Yard	Pennsylvania.
Miller, C. H	Nebraska.	Navy Yard	Virginia.
Miller, Rev. J	Kentucky.	Navy Yard	Tennessee.
Miller, John H	Kansas.	Neely, E. B	Missouri.
Miller, L. A	Vermont.	Nelson, H. M	Massachusetts.
Milliken, John T	Pennsylvania.	Nelson, Professor J. P	Maryland.
Mills, Dr. B. F	Wisconsin.	Nelson, R. J	Nova Scotia.
Mills, John H	North Carolina.	Nelson, S. A	Massachusetts.
Minich, J. B	Michigan.	Neuleton, A	Virginia.
Mitchell, Dr. R. W	Tennessee.	Newcomb, Guilford S	Massachusetts.
Mitchell, Dr. S. F	Michigan.	Newcomb, J. B	Illinois.
Mitchell, Hon. W	Massachusetts.	Newkirk, R. M	Indiana.
Moeller, G	Wisconsin.	Newton, Rev. Alfred	Ohio.
Moffatt, Professor A. G	Indian Territory.	Newton, John	Florida.
Moke, Dr. Jas. E	Texas.	Newton, W. H	Wisconsin.
Moore, Professor A	Mississippi.	Nichols, C. J	Maine.
Moore, Dr. Alex. P	Arkansas.	Nicholson, Rev. J. J	Alabama.
Moore, Asa P	Maine.	Niell, Thos	Ohio.
Moore, Mrs. Belle	Missouri.	Noll, Arthur B	New Jersey.
Moore, C. R	Virginia.	Normal School	Massachusetts.
Moore, David	Mississippi.	North, Dr. S. D	Alabama.
Moore, Dr. G. F	North Carolina.	Norton, J. H	New York.
Moore, J. A	North Carolina.	Norton, Prof. W. A	Delaware.

List of meteorological observers, &c.—Continued.

Name.	State.	Name.	State.
Norvell, F.	Dakota Territ'y.	Pearson, John	Florida.
Noarse, H. J.	Wisconsin.	Peck, Dr. John E.	Florida.
Nuttamer, Prof. Fra. J.	Virginia.	Peck, Dr. W. R.	Ohio.
Oakfield. C. F.	Kansas.	Peeler, David	Pennsylvania.
Oakley, Thomas	Mississippi.	Pret, Abraham S.	Massachusetts.
Observatory Harvard College	Massachusetts.	Prince, George S.	South Carolina.
O'Donoghue, John	Illinois.	Pendleton, Dr. E. M	Georgia.
Odell, Dr. B. F.	Minn. and Iowa.	Pendleton, P. C	Georgia.
Odell, Fletcher	New Hampshire.	Percival, Dr. Charles F.	Alabama.
Offutt, Dr. J. J.	Virginia.	Perkins, Capt. A. D	Michigan.
Oficinas Central de Estadistica	Costa Rica.	Perkins, Dr. H. C	Massachusetts.
Olds, Warren	Illinois.	Pernot, Claudius	New York.
Oliver, John	Michigan.	Perrault, Ed	Michigan.
Ormsby, G. S	Ohio.	Perry, Rev. J. B	Mass. and Vt.
Orton, James	Mass.	Perry, W	New Jersey.
Osborne, Dr. T. C	Alabama.	Peters, Adam	Ohio.
Osgood. H. H	Maine.	Peters, H	Indiana.
Owen, Benjamin	West Virginia.	Peters, Thos. M	Alabama.
Owen, Dr. D. D	Indiana.	Pettingill, Waldo.	Maine.
Owsley, J. B.	Ohio.	Pettingill, W	Ohio.
Packard, Levi S	New York.	Petty, Chas.	South Carolina.
*Paddock, James A	Vermont.	Petty, McK	Vermont.
Paine, Charles L	Vermont.	Phelps, H. E	Utah.
Paine, Charles S	Vermont.	Phelps, Highland W	Wisconsin.
Paine, Dr. H. M	New York.	Phelps, Rev. Joshua	Iowa.
Palm, Swante	Texas.	Phelps, R. H	Connecticut.
Palmer, Miss Jerusha B.	New Jersey.	Phelps, W. W	Utah.
Pardee, Eugene	Ohio.	Phillips, H	Canada.
Pardee, H. C	Nebraska.	Phillips, Prof. Jas., D. D.	North Carolina.
Park, William K	Virginia.	Phillips, J. H	Ohio.
Park, Rev. Roswell	Wisconsin.	Phillips, R. C	Ohio.
Parker, J. B	Michigan.	Pickard, Dr. J. L	Wisconsin.
Parker, J. D.	Maine.	Pickett, John	Virginia.
Parker, John D	Illinois.	Pierce, Warren	Ohio.
Parker, Joseph	Vermont.	Pillsbury. M. A	Ohio.
Parker, Dr. Joseph M	Tennessee.	Pillsbury, Mrs. M. A.	Ohio.
Parker, Melsar.	Wisconsin.	Pitcher, Dr. Zeno	Michigan.
Parker, Nathan H.	Iowa.	Pitman, Charles H	New Hampshire
Parker, Prof. W. H.	Vermont.	Pitman, E	Maine.
Parkinson, D. F.	California.	Pitman, M	Maine.
Parmelee, Ezra	New York.	Plumb, Dr. Ovid	Connecticut.
Parry, William	New Jersey.	Plummer, Dr. John T.	Indiana.
Parsons, B. W	Massachusetts.	Poe, James H	Ohio.
Parsons, L. H	New Jersey.	*Pollard, T. F	Vermont.
Partrick, J. M	New York.	Pollock, Rev. J. E	Missouri.
Parvin, P. O.	Iowa.	Pomeroy, F. C	Wisconsin.
Parvin, Prof. Theo. S	Iowa.	Pool, Isaac A	Illinois.
Pashley, J. S	Wisconsin.	Poole, H	Nova Scotia.
Paterson, Rev. Dr. A. B.	Minnesota.	Porter, E. D	Delaware.
Patrick, Dr. John J	Illinois.	Porter, Mrs. E. D.	Delaware.
Pattison, H. A	Michigan.	Porter, Henry D	Wisconsin.
Patton, Thomas	Virginia.	Porter, Prof. W	Wisconsin.
Patton, Dr. Wm. N	Virginia.	*Posey, Dr. John F.	Georgia.
Paxton, J. W	Michigan.	Post Surgeon	Kansas and Cal.
Payne, Dr. J. W	Alabama.	Potter, Dr. A. M	Texas.
Peabody, Prof. S. H	Vermont.	*Potter, C. D	New York.
Peale, Dr. J. B.	Pennsylvania.	Potter, G. W	New York.
Pearce, Harrison	Utah.	Potts, Jesse G	Virginia.
Pearce, James A., jr	Maryland.	Powel, Samuel	Rhode Island.
		Pratt, Prof. D. J	New York.
		Pratt, George B	Iowa.

List of meteorological observers, &c.—Continued.

Name.	State.	Name.	State.
Pratt, Dr. J. F	Maine.	Robinson, Rev. E. S	Mississippi.
Prentiss, Dr. H. C	Massachusetts.	Roby, Charles H	Virginia.
Prescott, Dr. W	New Hampshire.	Rockwell, Miss Charlotte.	Connecticut.
Preston, Rev. N. O	Kansas.	Rockwell, John A	Georgia.
Prince, Charles	Maine.	Redman, Samuel	Massachusetts.
Prince, G	Maine.	Rea, J. H	Illinois.
Purdie, John B	Virginia.	Roe, Rev. Sanford W	Con. and N. Y.
Purdet, E	Honduras.	Rordol, W. D	Virginia.
Purmort, N	New Hampshire.	Roffe, C. L	West Virginia.
Quincy, W. C	Virginia.	Rogers, A. P	Ohio.
Race, James A	Missouri.	Rogers, Francis M	New York.
Raffensperger, E. D	Ohio.	Rogers, J. S	Illinois.
Rain, John G	Nebraska.	Rogers, O. P	Illinois.
Ralston, Rev. J. G	Pennsylvania.	Roos, Charles	Minnesota.
Rambo, E. B	Indiana.	Root, Dr. Martin N	Mass. and N. H.
Randall, R. B	California.	Root, Professor O	New York.
Rankin, Colin	H. B. T. & Canada.	Ross, D. R	H. B. Territory.
		Rosseau, M. C	Idaho.
Rankin, D. M	Ohio.	Rossiter, Professor G. R	Virginia.
Rankin, James	Connecticut.	Rothrock, William	Tennessee.
Ranted, E. L	Louisiana.	Ray, G. P	Missouri.
Rawer, John Heyl	Pennsylvania.	Royal Engineers	Bermudas.
Ravenel, H. W	South Carolina.	Royal Engineers	Nova Scotia.
Ravenel, Thomas P	South Carolina.	Royal Gazette	Bermudas.
Ray, Dr. John D	Kentucky.	Rocker, D. H	Texas.
Ray, L. G	Kentucky.	Ruffin, David L	Virginia.
Rayal, James T	Texas.	Ruffin, Julian C	Virginia.
Raymond, George	Massachusetts.	Ruggles, Homer	Missouri.
Raymond, W. A	Michigan.	Russell, Cyrus H	New York.
Read, D. E	Neb. and Iowa.	Russell, O. F	Arkansas.
Reade, J. M	Tex. & N. Mex.	Rutherford, M	Texas.
Reasure, Dr. F. M	Michigan.	Ryerson, Dr. Thomas	New Jersey.
Redding, Thomas D	Indiana.	Salisbury, Elias O	New York.
Reed, Edwin C	New York.	Samms, Dr. C. C	Ohio.
Reed, Isaiah	Iowa.	Sampson, Alexander	W. Territory.
Reed, Lt. Jos	Pennsylvania.	Sanders, R. O	Virginia.
Reid, Jas. M	Georgia.	Sanford, S	Ohio.
Reld, Dr. Robert K	California.	Sanford, Dr. S. N	Ohio.
Reld, Peter	New York.	Sanger, Dr. W. W	New York.
Reynolds, Henry	Maine.	Sargent, John S	Massachusetts.
Reynolds, J	Arkansas.	Sartorius, Dr. Charles	Mexico.
Reynolds, Lauriston	Maine.	Sartwell, Dr. H. P	New York.
Reynolds, Orrin A	Massachusetts.	Saurman, John W	Pennsylvania.
Reynolds, R. M	Alabama.	Savage, Dr. G. S	Kentucky.
Reynolds, W	Iowa.	Savery, Thomas H	Pennsylvania.
Reynolds, W. C	Virginia.	Saville, J. J	Iowa.
Rhees, Dr. Morgan J	New Jersey.	Sawkins, James G	Jamaica.
Rhoades, Dr. John	Ohio.	Sawyer, George D	N. Hampshire.
Rhodes, W. H. T	Texas.	Sawyer, H. E	N. Hampshire.
Riblet, J. H	Illinois.	Sawyer, Thomas L	Tennessee.
Rice, Frank H	Massachusetts.	Scudlin, Rev. Wm. G	Massachusetts.
Rice, Henry	Massachusetts.	Scarritt, Rev. N	Missouri.
Richards, Thomas	Il. B. Ter'y.	Schauber, H. A	New York, Ohio, and Illinois.
Rickmeker, L. E	Pennsylvania.		
Riddell, Prof. W. P	Louisiana.	Scheeps, E. H. A	Iowa.
Riggs, A. L	Minnesota.	Schenck, Dr. W. L	Ohio.
Riggs, Rev. S. R	Minnesota.	Schotterly, Dr. H. R	Michigan.
Riker, Walter H	New York.	Schlegel, Albert	Massachusetts.
Riotte, Hon. C. N	Costa Rica.	Schley, W	Georgia.
Ritchie, James	Massachusetts.	Schmidt, Dr. E. R	New Jersey.
Riter, F. G	Dakota.	Scholsfield N	Connecticut.
Robbins, Dr. James	Massachusetts.	Schriever, Francis	Pennsylvania.
Roberts, D. H	Indiana.	Schus, Dr. A	Wisconsin.
Robinson, Almon	Maine.	Scott, H. B	Florida.

List of meteorological observers, &c.—Continued.

Name.	State.	Name.	State.
Scott, James	Kansas.	Smith, C. B	Nebraska.
Scott, Samuel	Pennsylvania.	Smith, Dr. C. C	Michigan.
Scovill, H. W	Illinois.	Smith, C. E	Illinois.
Scriba, Victor	Pennsylvania.	Smith, Dr. Carl H	Ohio.
Seabrook, Thomas	Pennsylvania.	Smith, Eli	Maryland.
Seavey, C. C	Georgia.	Smith, Dr. E. A	Massachusetts.
Seibert, Samuel K	Wisconsin.	Smith, E. A, and daughters	New York.
Selby, Henry	Michigan.	Smith, E. L	Massachusetts.
Seltz, Charles	Nebraska.	Smith, Rev. George N	Michigan.
Senior class, Mount Au-		Smith, Dr. George O	Illinois.
burn Female Institute	Ohio.	Smith, Haden Patrick	New York.
Sergeant, John T	New Jersey.	Smith, Hamilton, Jr	Indiana.
Seymour, E	Wisconsin.	Smith, Henry K	Illinois.
Seymour, Dr. E. W	Kansas.	Smith, H. I	Minnesota.
Shackelford, Prof. J	Alabama.	Smith, Harmon M	Michigan.
Shaffer, J. M	Iowa.	Smith, Howard D	Maine.
Shane, J. D	Kentucky.	Smith, Isaac H	Illinois.
Sharp, Dr. W. H	West Virginia.	Smith, Dr. J. Bryant	North Carolina.
Shaw, Francis	Massachusetts.	Smith, J. C	Ohio.
Shaw, Jos	Ohio.	Smith, J. Edwards	Mississippi.
Shaw, M	Kansas.	Smith, John M	Missouri.
Sheerer, H. M	New York.	Smith, J. Metcalf	New York.
Sheldon, Daniel	Iowa.	Smith, Dr. J. W	New York.
Sheldon, D. S	Iowa.	Smith, Rev. L. M. S	Michigan.
Sheldon, H. A	Vermont.	Smith, M. D	California.
Sheldon, H. C	Rhode Island.	Smith, Mrs. M. D	California.
Shephard, J. A	Mississippi.	Smith, Dr. N. D	Arkansas.
Shepherd, Smiley	Illinois.	Smith, Rufus	New Hampshire
Sheppard, Benjamin	New Jersey.	Smith, Sydney	Iowa.
Sheppard, Clarkson	New Jersey.	Smith, Rev. S. U	Alabama.
Sheppard, Rev. J. A	N. C. and Ala.	Smith, Rev. W	Pennsylvania.
Sheppard, Miss R. C	New Jersey.	Smurr, Dr. T. A	Ohio.
Sherman, Rev. D. H	Illinois.	Snyser, Dr. B. R	Pennsylvania.
Shields, J. H	Alabama.	Snyser, Dr. H	Pennsylvania.
Shields, Rev. R	Ohio.	Snell, Prof. F. S	Massachusetts.
Shiniz, H. J	Wisconsin.	Snow, Prof. F. H	Kansas.
Shoemaker, J. G	Kansas.	Snyder, James A	Oregon.
Shortwell, D. F	Minnesota.	Soule, Prof. W	New York.
Shotwell, Samuel L	Illinois.	Soule, W. L. G	Kansas.
Shreve, Charles R	Ohio.	Southworth, N. C	Michigan.
Shriver, Howard	Va. and N. J.	Spaulding, Dr. Abiram	Illinois.
Shuman, Bruno	Texas.	Spaulding, S. C	Illinois.
Shumard, G. G	Texas.	Spates, Samuel	Minnesota.
Sias, Professor Solomon	Tex. and N. Y.	Sprer, Dr. Alex M	Pennsylvania.
Siber, Audr. L	Utah.	Spencer, Miss Anna	Pennsylvania.
Sibley, P. D	Missouri.	Spencer, Edward E	Virginia.
Simpson, D. F	New Jersey.	Spencer, Rev. D. D	Minnesota.
Simpson, F. T	Georgia.	Spencer, E. S	Wisconsin.
Simson, Rodman	Pennsylvania.	Spencer, W. C	Illinois.
Skeen, W	Virginia.	Spera, W. H	Pennsylvania.
Slade, Frederick J	New York.	Sperry, M	Ohio.
Slason, I. H	California.	Spencer, F	Ohio.

List of meteorological observers, &c.—Continued.

Name.	State.	Name.	State.
Stebbins, Richard	Iowa.	Taylor, Dr. M. K	Michigan.
Steed, F.	Iowa.	Taylor, Rev. R. T	Pennsylvania.
Steele, Hon. Aug	Florida.	Teele, Rev. A. K	La. and Mass.
Steele, George E	Michigan.	Tenbrock, J. W	Indiana.
Steiner, Dr. Lewis F	Maryland.	Terry, Charles C	Massachusetts.
Stephens, Prof. A. M	Minnesota.	Tew, Captain C. C	South Carolina.
Stephens, J. A	Missouri.	Thatcher, A. E	New York.
Stephenson, Rev. James	Maryland.	Thickstan, T. F	Minnesota.
Sterling, J. W	Wisconsin.	Thickstan, T. F	Pennsylvania.
Stern, Jacob T	Iowa.	Thomas, Mrs. W. S	Illinois.
Sternbergh, W. H	New Granada.	Thompson, A. H	Illinois.
Stevens, Hennell	Texas.	Thompson, Rev. D	Ohio.
Stevens, Dr. J. I.	Maine.	Thompson, D. P	Vermont.
Stevens, Linus	New Hampshire.	Thompson, Rev. E	Ohio.
Stevens, R. P	Pennsylvania.	Thompson, E. J'	Illinois.
Stewart, Prof. A. P	Tennessee.	Thompson, George W	New Jersey.
Stewart, F. L	Pennsylvania.	Thompson, H	Louisiana.
Stewart, Thomas H	Pennsylvania.	Thompson, Prof. H. A.	Ohio.
Stewart, W. M	Tennessee.	Thompson, Mrs. Phœbe	Vermont.
Stockwell, George A	Michigan.	Thompson, R. O	Nebraska.
Stokes, H. A	New Jersey.	Thompson, Prof. Zadok	Vermont.
Stokes, William A	Pennsylvania.	Thomson, Prof. S. H	Indiana.
Stone, Isaac	Michigan.	Thornton, Miss E. E	New Jersey.
Stonffer, And	Minnesota.	Thornton, Dr. S. C	New Jersey.
Stowell, T. D	Kansas.	Thorp, Henry W	Md. and Penn.
Stractmans, H. I	Missouri.	Tidswell, Miss Mary A	Missouri.
Strang, J. J	Michigan.	Tingley, Prof. Joseph	Indiana.
Streng, L. H	Michigan.	Tirrell, Dr. N. Q	Massachusetts.
Strickland, L. S	Maine.	Titcomb, J. S	Illinois.
Strong, Edwin A	Michigan.	Titus, H. W	New York.
Strong, Oscar I	Iowa.	Tize, H. A	Illinois.
Strong, Rev. Thomas H	New York.	Toby, James K	Vermont.
Strunk, Daniel	Wisconsin.	Tolman, J. W	Illinois.
Struthers, R. H	Wisconsin.	Tolman, Rev. Marcus A	Pennsylvania.
Stuart, E. W	Ohio.	Tompkins, W	New York.
Stuart, Prof. A. P. S	Nova Scotia.	Tooker, Nathan C	Pennsylvania.
Stumps, S. J	West Virginia.	Toothellot, Dr. I.. A	New York.
Sunstebeck, F. H	Missouri.	Tower, James M	New York.
Stuots, G. R	Wisconsin.	Towers, M. H	Wisconsin.
Struver, A. S	Ohio.	Townsend, Nathan	Iowa.
Suter, Captain C. R	South Carolina.	Towson, J. W	Ohio.
Sutherland, Norris	Missouri.	Tracy, George H	Pennsylvania.
Sutton, Rev. A	Maryland.	Tracy, James F	Pennsylvania.
Sutton, Dr. G	Indiana.	Travelli, J. I	Pennsylvania.
Swain, Dr. John	Illinois and Ky.	Treat, Samuel W	Ohio.
Swan, Caleb	New York.	Trembley, Dr. J. B	Ohio.
Swan, James G	Washington Ter.	Trevor, James B	New York.
Swart, Haren V	New York.	Trible, Miss Anna C	Illinois.
Swazey, C. D	La. and Miss.	Trivett, Walter M	California.
Swift, Lewis	New York.	Trowbridge, David	New York.
Swift, Dr. Paul	Pennsylvania.	Troy, Dr. M	Alabama.
Sylvester, Dr. E. Ware	New York.	True, H. A	Ohio.
Talcott, H	Illinois.	' Tuck, Dr. W. J	Tennessee.
Tappan, Eugene	Massachusetts.	Tucker, Edward T	Massachusetts.
Tappan, E. T	Ohio.	Tuckerman, L. B	Ohio.
Tare, And	Wisconsin.	Tuomey, Prof. M	Alabama.
Tavel, B. F	Tennessee.	Turnbull, Lieut. C. N	Michigan.
Taylor, E. T	Virginia.	Turner, A. P	Indiana.
Taylor, John	Pennsylvania.	Turner, David	Virginia.
Taylor, Joseph W	New York.	Turner, T. A	Texas.
Taylor, Prof. K. M	Iowa.	Tutwiler, H	Alabama.
Taylor, L. B	Louisiana.	Tweedy, D. H	Ohio.

List of meteorological observers, &c.—Continued.

Name.	State.	Name.	State.
Twiss, Thomas S	Nebraska.	Wattles, J. O	Kansas.
United States consul	Bahamas.	Weast, J. W	Arkansas.
Ufford, Rev. John	Iowa.	Weatherford, John M	Missouri.
Underwood, D	Vermont.	Webb, Miss G	Mich. and Ind.
Underwood, Colonel D	Wisconsin.	Webb, Dr. Robert D	Alabama.
Upshaw, G. W	Virginia.	Weber, P	Missouri.
United States engineers	Michigan.	Webster, Prof. N. D	Virginia & N. C.
Vagnier, Thomas	Indiana.	Weeks, James A	Pennsylvania.
Valentine, John	Indiana.	Weir, A. D	Pennsylvania.
Valentine, Felipe	Costa Rica.	Weiser, R	Pennsylvania.
Van Blascom, J	Maine.	Wellford, B. R	Virginia.
Van Buren, Jarvis	Georgia.	Wells, C. D	Nebraska.
Van Doren, A	Virginia.	Wells, J. Carson	Virginia.
Van Horne, F. D	Indian Territory.	Wells, W	Missouri.
Vankirkle, L	Delaware.	West, Edmund	Ohio.
Vankirk, W. J	Alabama.	West, E. W	Ohio.
Vankirk, W. J	Missouri.	West, L. W	Wisconsin.
Vankleek, Rev. R. D	New York.	West, Dr. N. P	Texas.
Van Nostrand, J	Texas.	West, Silas	Maine.
Van Orden, W	Michigan.	Westbrook, Samuel W	North Carolina.
Van Vorhees, A	Minnesota.	Westdahl, F	Alaska Ter.
Von Frantzius, Dr. A	Costa Rica.	Westmoreland, J. G	Georgia.
Verrill, G. W., jr	Maine.	Wetherill, Prof	New York.
Vertrees, John E	Missouri.	Wheaton, Alex. Camp	Iowa & Mon. T.
Vincent, J. H	Mississippi.	Wheaton, Mrs. Daniel D	Iowa.
Vogel, C	Missouri.	Wheeler, B. J	Vermont.
Waddell, William H	Mississippi.	Wheeler, John T	New Hampshire.
Wade, Edward	Ohio.	Whelpley, Miss Flor'e E	Michigan.
Wade, F. S	Texas.	Whelpley, Miss Helen I	Michigan.
Wadey, H	Iowa.	Whelpley, Dr. Thomas	Michigan.
Wadsworth, A. S	New York.	Whipple, Capt. A. W	Michigan.
Wadsworth, General P	Maine.	Whitaker, D	Illinois.
Wagner, W. H	Oregon.	Whitcomb, George	Missouri.
Wainwright, Elmore	Michigan.	Whitcomb, L. F	Massachusetts.
Waite, M. C	Wisconsin.	White, Prof. Aaron	New York.
Waksoley, C. C	New York.	White, Bela	Nebraska.
Walker, David, M. D	Vancouver's Isl.	White, Prof. J. D	South Carolina.
Walker, J. P	Delaware.	White, Peter	Michigan.
Walker, Mrs. Mary A	Kentucky.	White, Dr. W. T	New Grenada.
Walker, Mrs. Octavia C	Michigan.	Whitehead, W. A	New Jersey.
Walker, H. L	Pennsylvania.	Whitfield, E	Minnesota.
Walker, S. C	Pennsylvania.	Whiting, Robert C	New Hampshire.
Wallace, Samuel J	Illinois.	Whiting, William H	Wisconsin.
Wallace, Colonel W	South Carolina.	Whitlock, James H	California.
Waller, R. B	Alabama.	Whitner, B F	Florida.
Walsh, Stephen	Minnesota.	Whitney, Miss L. J	Georgia.
Walter, Dr. James	Kansas.	Whittlesey, C. S	Michigan.
Walton, Joseph P	Iowa.	Whittlesey, S. H	Michigan.
Ward, Rev. L. F	Ohio.	Wichline, Thomas J	Virginia.
Ward, Prof. W. H	Wisconsin.	Wieland, C	Minnesota.
Warder, A. A	Ohio.	Wieland, H	Minnesota.
Warder, R. D	Ohio.	Wiesner, J	Dist. Columbia.
Waring, Prof. C. D	New York.	Wiggin, Andrew	New Hampshire.
Warne, Dr. George	Iowa.	Wilbur, B. F	Maine.
Warren, James H	New York.	Wild, Rev. Edward P	Massachusetts.
Warren, James H	Iowa.	Wilkinson, C. H., M. D	Texas.
Washburn, D	Pennsylvania.	Wilkinson, John H	Ohio.
Washington, L	Wisconsin.	*Willard, J. F	Wisconsin.
Watkins, W. D	Ohio.	Willet, Prof. J. E	Georgia.
Watson, George	New Jersey.	Williams, H. C	Illinois.
Watterson, H R	Ohio.	Williams, Ed. F	Tennessee.
Wattles, Miss Celestia	Kansas.	Williams, H. B	Iowa.

List of meteorological observers, &c.—Continued.

Name.	State.	Name.	State.
Williams, Prof. J. R.	Pennsylvania.	Woodruff, E. N.	Kentucky.
Williams, Prof. I. D.	Pennsylvania.	Woodruff, L.	Michigan.
Williams, Prof. M. G.	Ky. & Ohio.	Woods, W.	Wisconsin.
Williams, Dr. P. O.	New York.	Woodard, C. S.	Michigan & Ind.
Williams, Rev. R. G.	N. Y. & Conn.	Woodard, C. S.	New York.
Williams, Rev. S. R.	Kentucky.	Woodward, Lewis	New York.
Williams, Prof. Wm. D.	Georgia.	Woodworth, Dr. A.	Kansas.
Willis, Henry	Maine.	Woodworth, Samuel	Iowa.
Willis, O. R.	N. Y. & N. J.	Woolsey, Dr. W. W.	Iowa.
Willis, P. L.	Oregon.	Wooster. C. A.	New York.
Wills, J. C	Virginia.	Wormley, Theo. G.	Ohio.
Wilson, G. W., jr	Missouri.	Wray, Alex	North Carolina.
Wilson, Joseph A.	Missouri.	Wright, Dr. Daniel F.	Tennessee.
Wilson, Dr. J. B.	Maine.	Wright, E. M.	Minnesota.
Wilson, Prof. J. H.	Ohio.	Wyman, A. H.	Maine.
Wilson, Lavallette	Massachusetts.	Wyrick, M. L.	Missouri.
Wilson, P. S.	Missouri.	Yale, Walter D.	New York.
Wilson, Prof. W. C.	Pennsylvania.	Yellowby, Prof. C. W.	Texas.
Wilson, Rev. W. D.	New York.	Yeomans, W. G.	Connecticut.
Wilson, W. W.	Pennsylvania.	Yoakum, F. L.	Texas.
Winebell, Prof. A.	Ala. & Michigan.	Yoakum, H.	Texas.
Windle, Isaac E.	Indiana.	Young, A. A.	New Hampshire.
Wing, M. E.	Vermont.	Young, Prof. C. A.	Ohio.
Winger, Martin	Ohio.	Young, Prof. Ira	New Hampshire.
Winkler, Dr. C.	Wisconsin.	Young, J. A.	South Carolina.
Wise, John	Pennsylvania.	Young, Judo. M.	New York.
Wislizenus, Dr. A.	Missouri.	Young, Mrs. L.	Kentucky.
Withrow, Thomas F.	Ohio.	Younger, Armistead	Arkansas.
Witter, D. K.	Iowa.	Younglove, J. E.	Kentucky.
Woodbridge, W.	Indiana.	Zaeppfel, J.	New York.
Woodbury, C. F.	Minnesota.	Zahnor, P.	Nebraska.
Woodbury, C. W.	Minnesota.	Zealy, Joseph T.	South Carolina.
Wood, Samuel	New Jersey.	Zeigler, Dr. A. F.	Idaho Territory.
Wood, S. D.	Ohio.	Zimmerman, G.	New York.
Woodbridge, W.	Indiana.	Zombrock, Dr. A.	Maryland.
Woodin, S. F.	New York.		

METEOROLOGICAL MATERIAL CONTRIBUTED IN ADDITION TO THE REG-
ULAR OBSERVATIONS, DURING THE YEAR 1868.

Académie Impériale de Lyon.—Mémoires de l'Académie Impériale des
Sciences, Belles-Lettres et Arts de Lyon, classe des sciences, tome seizième,
Lyon, 1866–'67, 8vo., 416 pages. [Contains: Observations Météorologiques faites
à 9 heures du matin à l'observatoire de Lyon du 1er Décembre 1865, au 1er
Décembre 1866, par M. Aimé Drian, sous la direction de M. Lafon, professeur
à la Faculté des Sciences et directeur de l'observatoire. 26 pp. Résultats de la
nouvelle série d'observations ozonométriques faites par MM. le Docteur Lembert
et F. Rassinier, durant l'année 1866, 32 pp. Classification des phénomènes
produits par l'électricité météorique dans le bassin du Rhône et aux alentours,
par M. J. Fournet, président de la commission des Orages, lue dans la séance
du 6 Mars 1867, 122 pp.]

Aguilar, F. C.—Boletino Meteorologico del observatorio del Colegio Nacional
de Quito, dirigido por los padres de la Compañia de Jesus, segundo año 1866,
F. C. Aguilar, S. J., Quito, 1868, 36 pages, 8vo.

Asiatic Society of Bengal.—Journal, Part II, No. II, 1868, containing abstract
of the results of meteorological observations taken at the surveyor general's office,
Calcutta, September, 1866, to June, 1867, inclusive; also tables of mean
monthly readings of the barometer reduced to freezing point, for 10 years, from
1856 to 1865, and of barometric curve and registered rainfall.

Bache, R. M.—Notes on the climate of San Francisco, California, and table
of temperature.

Berkey, W. H.—Meteorological record kept at Ossawatomie, Kansas, during
the month of March, 1868.

Boardman, G. A.—Meteorological observations made during the month of
January, 1868, at Green Cove Spring, Florida.

Boerner, Charles G.—Account of the meteoric shower of November 13 and
14, observed at Vevay, Indiana.

Bruhns, Dr. C.—Resultate aus den Meteorologischen Beobachtungen angestellt in mehreren Orten im Königreich Sachsen in den Jahren 1826 bis
1851, und an den fünfundzwanzig Königlichen Sächsischen Stationen im Jahre
1866. Nach den monatlichen Zusammenstellungen im Statistischen Bureau des
Königlichen Ministeriums des Innern, bearbeitet von Dr. C. Bruhns, Director
der Sternwarte und Professor der Astronomie in Leipzig. Dritter Jahrgang,
Leipzig, 1868, 4to, 136 pp.

Bush, Richard J.—Meteorological observations made during the exploration
of the country between Okhotsk and Nicolaefsk, East Siberia, in the service of
the Western Union Telegraph Company, October 21, 1865, to March 4, 1866.

Chase, Pliny Earle, Philadelphia.—Curves indicating the relation of the moon
to temperature at the surface of the earth.

Christiania Observatorium.—Meteorologiske Iakttagelser, 1867.

Cornelissen, J. D.—On the temperature of the sea near the south point of
Africa. Royal Meteorological Institute of the Netherlands, 4to.

Cox, Judge Hopewell.—Meteorological observations made at Hartford, Dodge
county, Wisconsin, from April, 1859, to September, 1862; copied by S. G.
Lapham, for the Smithsonian Institution.

Dall, W. H.—Meteorological observations made at Nulato, Alaska, from
December 1, 1866, to May 26, 1867, by W. H. Dall; also meteorological obser-
vations at Unalakleet, Alaska, from October 19, 1866, to January 23, 1867, by
F. Westdahl.

Davis, R. J.—Observations of temperature near Glasgow Station, Amherst
county, Virginia.

Destruge, A.—Meteorological register for the month of October, 1868, at

Draper, Dr. Joseph.—Meteorological observations made at the State Lunatic Hospital, Worcester, Massachusetts, during the years 1859, 1860, and 1861.

Engineer corps, United States army.—Hourly curves at Willet's Point, New York, from 12 days' hourly observations, July 14 to July 25, 1868; 8vo, 2 pp.

French, J. B., agent W. Lake C. & W. Manufacturing Company, Lake Village, New Hampshire.—Tables of rainfall at Lake Village and Laconia, New Hampshire.

Galveston City Hospital.—Meteorological observations made at the Galveston City Hospital, Galveston, Texas, during the year 1867, by Drs. C. H. Wilkinson, H. A. McComly, and others.

Gesellschaft für Natur-und Heilkunde in Dresden.—Die Vertheilungen der Windstärke in der Windrose von Dresden, von Dr. Ed. Zeisler; 8vo.

Gibbs, George.—Meteorological statement for the year ending October 31, 1868, Sitka, Alaska Territory.

Gilman, W. S., Jr.—Weather items, Palisades, Rockland county, New York, 1868.

Hann, Dr. Julius.—Die thermischen Verhältnisse der Luftströmungen auf dem Obir (6,288 Par. Fuss) in Kärnthen, mit 1 Tafel; December, 1867.

Die Temperatur-Abnahme mit der Höhe als ein Function der Windesrichtung, mit 1 Tafel; March, 1868.

Zur Charakteristik der Winde des adriatischen Meers, mit 1 Tafel; 1868.

Hart, Charles H.—Meteorological register kept at Paraná, South America, from October, 1843, to July, 1850, inclusive.

Heis, Dr.—Wochenschrift für Astronomie, Meteorologie und Geographie. Neue Folge, Elfter Jahrgang; (der "Astronomischen Unterhaltungen," 22ter Jahrgang.) Redigirt von Professor Dr. Heis, in Münster, 1868; (8vo, each number eight pages.)

Hoff, Alexander H., Assistant Surgeon United States Army.—Meteorological statement for the year ending October 31, 1868. Sitka, Alaska Territory.

Huntingdon, George C.—Meteorological tables for Kelley's island, Ohio, compiled from ten years' observations, 1859 to 1868 inclusive. Newspaper slip.

Third annual report of the Lake Shore Grape-growers' Association, containing article on climatology and grape culture, by George C. Huntington. 8vo.

Institut Egyptien.—Mémoires ou travaux originaux presentés et lus à l'Institut Égyptien, publiés sous les auspices de S. A. Mahommed-Saïd, Vice-Roi d'Egypte, sous la direction de M. le Docteur B. Schnepp, Secrétaire de l'Institut Égyptien. Tome premier. Paris, 1862, 4to, 776 pages. [Contains: Études sur le climat de l'Egypte, par le Docteur B. Schnepp, 192 pages. Du Khamsin et de ses effets; du Blé rotrait, par M. Grégoire, membre de l'Institut Égyptien, 14 pages.] (The *Khamsin* is a desert wind called elsewhere Simoon.)

Jelinek, C.—Zeitschrift der österreichischen Gesellschaft für Meteorologie. Redigirt von C. Jelinek und J. Hann. 8vo, Vienna, published twice a month

Jourdan, C. H., Prof.—Meteorological curves and summary of observations for 1868, at Mount St. Mary's College, Emmettsburg, Maryland.

Kaiserlich-Königliche Central-Anstalt für Meteorologie und Erdmagnetismus.—Jahrbücher der K. K. Central-Anstalt für Meteorologie und Erdmagnetismus, vor Carl Jelinek, Director, und Carl Fritsch, Vice-Director. Neue Folge, II Band Jahrgang 1865. Der ganzen Reihe XTer Band. Wien 1867, 4to, 212 pages Jahrbücher, Wien. 1868, 4to, 205 pp.

Kluge, J. P., M. D.—Meteorological observations made at Aspinwall, U. S C., from the week ending January 18, 1868, to week ending January 2, 1869

Koninklijk Nederlandsch Meteorolog. Instituut.—Jaarboek voor 1867. Parte 1 and 2, oblong 4to, pp. 244 and 114.

Koninklijk Nederlandsch Meteorologisch Instituut.—Meteorologische Waarnemingin in Nederland en zijne bezittingen, en Afwijkingen van Temperatuur en Barometerstand op vele Plaatsen in Europa. Uitgegeven door het Koninklijk

Nederlandsch Meteorologisch Instituut. 1864. Utrecht, 1865. Oblong 4to, 308 pages.

Lewis, Charles H.—Thermometrical record at Elizabethton, Carter county, East Tennessee, for the month of February, 1868.

Lewis, Dr. James.—Hourly records for the year 1868, by his self-recording barometer and thermometer, and reductions of the same, at Mohawk, New York.

Little, Frank.—Thermometrical observations at Kalamazoo, Michigan, during May, June, and July, 1868

Livings, B. C—Meteorological register at Smithfield, Wabash county, Minnesota, during the month of January, 1868.

Lupton, N. T.—Abstract of meteorological register for 1868, at the Southern University, Greensboro, Alabama. (Newspaper slip.)

Macgregor, O. J., M. A.—Abstract of meteorological observations at Stratford, Canada, for the year 1867. (Newspaper slip.)

Mackey, Robert.—Thermometrical record for July, 1868, at Island creek, Jefferson county, Ohio.

Magnetic Observatory, Toronto, Canada.—General meteorological registers for the year 1867. 8vo. 6 pp.

Marsh, Roswell.—Summary of meteorological observations during the year 1868. at Steubenville, Ohio.

Merriam, C. C.—Meteorological report for the year 1868, at Locust Grove, Lewis county, New York.

Meteorological Society, (British.)—Proceedings of the Meteorological Society, edited by James Glaisher, esq, F. R. S., president. 8vo. London, monthly. Meteorology of England. By James Glaisher. Quarterly.

Meteorologische Centralanstalt der Schweizerische Naturforschende Gesellschaft—Meteorolog. Beobachtungen, 1867, and January and February, 1868.

Michigan Board of Agriculture.—Report for 1867, containing a paper on the influence of forest trees on agriculture, and a meteorological register for 1867, by Professor R. C. Kedzie, State Agricultural College. Lansing.

Miller, Lester A.—Maxima and minima of temperature at Woodstock, Vermont, during the month of November, 1867.

Moore, C. R.—Thermometrical record for the month of March, 1868, at Bridgeton post office, near Eastville, Northampton county, Virginia.

Moultrie, J. L.—Summary of rainfall at Union Springs, Alabama, during 1868.

National Military School of Medicine of Roumania, [through the United States Naval Observatory, Washington.] Meteorological tables.

Naturaliste Canadien.—[Contains: Meteorological register at Port Neuf, Canada, for 1868.]

Naturforschende Gesellschaft su Görlitz.—Abhandlungen der Naturforschenden Gesellschaft zu Görlitz. Dreizehnter Band. Görlitz, 1868, 8vo, 296 pp. [Contains :- Meteorologische Beobachtungen in Görlitz vom 1 December, 1863, bis 30 November, 1866, von R. Peck. P. 125-208.]

Norske Meteorologiske Institut.—Norsk meteorologisk aarbog for 1867.

Observatory of Upsala.—Observations météorologiques faites à l'observatoire d' Upsala pendant les années 1855 et 1861.

Observatoire Royal de Bruxelles.—Annales météorologiques do l'observatoire Royal de Bruxelles, publiées, aux frais de l'état, par le directeur A. Quetelet, Première Année. Bruxelles, 1868, 4to, 96 pp

Palm, Swante.—Meteorological registers kept at Austin, Texas, from 1860 to 1864, inclusive.

Parvin, Prof. T. S.—Summary of meteorological observations at Iowa City, Iowa, during the year 1868. 8vo, 2 pp.

Pattison, H. A.—Report of meteorological observations at Muskegon, Michi gau, during part of the month of August, 1868.

Physikalischer Verein.—Jahresbericht des Physikalischen Vereins zu Frank furt am Main, für das Rechnngsjahr 1866-'67, 8vo., 120 pages. [Contains: Vermischte Meteorologische Notizen, von Professor Dr. Oppel, 19 pp. Ueber tägliche Barometerschwankungen und das Gesetz der täglichen Drehung des Windes, von Dr. Berger, 20 pp. Meteorologische Notizen vom Jahre 1867, 7 pp. Wasserhöhe des Mains vom Jahre 1867, 1 page. Gewonnene Ergeb nisse aus den im Jahre 1867 angestellten meteorologischen Beobachtungen des Physikalischen Vereins, 4 pp. Graphische Witterungstabelle des Jahres 1867.]

Poey, M. André.—Bibliographie Cyclonique, deuxième edition, Paris, 1866.

"Generalités sur le climat du Mexique, et sur l'éclipse totale de lune du 30 Mars dernier," Paris, August, 1866.

"Sur l'inversion diurne et nocturne de la température jusqu'aux limites de l'atmosphère, et à sa repartition de l'horizon au zénith," January, 1865.

"Sur la non-existence sous le ciel du Mexique, de la grande pluie d'étoiles filantes de Novembre, 1866, et du retour périodique du mois d'Août," 1867.

"Sur la non-existence sous le ciel austral des retours périodiques des étoiles filantes, et sur leur extinction graduelle du pole nord à l'équateur," October, 1865.

"Description d'un ozonographe et d'un actinographe destinés à enregistrer, do demi-heure à demi-heure l'ozone atmosphérique, et l'action chimique de la lumière ambiante," December, 1865.

"Remarques sur les colorations ozonoscopiques obtenues à l'aide du réactif de Jaine (de Sedan,) et sur l'échelle ozonométrique de M. Derigny," October, 1867.

"Travaux sur la météorologie, la physique du globe en general, et sur la clima tologie de l'ile de Cuba et des Antilles," October, 1861.

Appel aux nations Hispano-Americaines. (Meteorological circular.)

Pratt, W. H.—Account of meteoric shower observed on the night of Novem ber 13 and 14, 1868, at Davenport, Iowa.

Radcliffe Observatory.—Results of astronomical and meteorological observa tions made at the Radcliffe Observatory, 1865.

Ravenel, T. P.—Meteorological Journal for the year 1860, kept at St. John's, Berkeley parish, South Carolina, for the Black Oak Agricultural Society, by T. P. Ravenel, secretary. Pamphlet, 8vo, 15 pp. Charleston, 1861.

Ray, Dr. John D.—Observations at Paris, Kentucky, during January to July, 1855.

Real Academia de Ciencias, &c, Habana.—Anales. (Containing meteorolog ical registers.)

Real Observatorio de Madrid.—Observaciones Meteorológicas effectuadas en el Real Observatorio de Madrid, desde 1° de Diciembre de 1865, al 30 de Noviem bre de 1866. Madrid, 1867. 16mo. 175 pp.

Informe del Director del Real Observatorio astronomico y meteorológico de Madrid al Excmo. Sr. Comisario Regio del mismo establecimiento. Madrid, 1867. 16mo. 72 pp.

Resumen de las Observaciones Meteorologicas efectuadas en la Peninsula desde el 1° de Diciembre de 1865, al 30 de Noviembre de 1866. Madrid, 1867. 16mo. 364 pp.

R. Osservatorio di Palermo.—Bulletino Meteorologico del R. Osservatorio di Palermo. (Estratto dal Giorn. di Scienze Naturali ed Econ.) 4to, monthly.

Rijks Observatorium, Leiden.—Onderzoekingen ombrent den Gang van het Hoofduurwerk des Sternwacht to Leiden de Pendule. Nobwii No. 17. Door F. Kaiser.

Rocke, Edouard.—Recherches sur les Offuscations du Soleil et les Météores Cosmiques. Paris, 1868.

Royal Agricultural Society of England.—The Journal of the Royal Agricul tural Society of England. Second series, Volume IV. London, 1868. 8vo,

294 pages [Contains: Meteorology for the six months ending December 31, 1867—5 pp. On the Temperature of the Sea and Its Influence on the Climate and Agriculture of the British Isles; by Nicholas Whitley, F. M. S.—31 pp.

Ruffner, W. H.—Weather notes taken at Tribrook farm, one and a half mile southwest of Lexington, Virginia, during the years 1867 and 1868.

Sawyer, Henry, United States consul.—Meteorological register at Paramaribo, Dutch Guiana, from January 12, 1868, to January 3, 1869. (Newspaper slips.)

Scdei Nationale de Medicina.—Tablou General de Observatiunele Meteori ologice ale Scolei Nationale de Medicina facute la Spitalul militar din Bucuresti in Anul 1863. Directorul scholei, Davila, observator; A. Lessman. (One large lithograph sheet of tables and diagrams. Similar sheets for 1864, 1865, 1866, 1867.)

Scottish Meteorological Society.—Journal of the Scottish Meteorological Society. Published quarterly. 8vo. Edinburgh.

Scudder, S. H.—Meteorological observations taken during a trip to Cuba and the Isle of Pines in the spring of 1864.

Shepherd, Smiley.—Abstract of observations for each month of the year 1868, at Hennepin, Illinois.

Sisson, Rodman.—A table showing the temperature of the seasons for the three years from December, 1864, to November, 1867. inclusive, at North Abingdon, Luzerne county, Pennsylvania.

Swiheart, Dr. C. F.—Meteorological observations made at Houston, Texas, by Dr. A. M. Potter, from September, 1862, to October, 1865. (Dr. Potter died October 10, 1865, and this register was presented to Dr. Swiheart by his family.)

Société Météorologique de France.—Nouvelles Météorologiques, publiées sous les auspices de la Société Météorologique de France. 8vo. Paris. (Issued monthly.) Annuaire 1866, 1867.

Società Reale di Napoli.—Rendiconto dell' Academia delle Scienze Fisiche e Matematiche. April, 1865, to May, 1867. Quarto, monthly. [April, 1865, contains: "Nuovo anemografo elettro-magnetico. Memorie del Socio Ordinario L. Palmieri;" pp. 105–106. June, 1865: "Tremblement de terre de 1862 dans les environs du Lac Baikal, (Siberie Orientale,) par P. Kropotkine et A. Palibine;" pp. 181–193. August, 1865: "Se le osservazioni di meteorologica elettrica fatte con l'elettrometro bifilare e col conduttore mobile siano comparabili; nota del Socio Ordinario L. Palmieri;" pp. 244–249. September, 1865: "Procella magnetica contemporanea all' apparizione delle stelle cadenti; nota del L. Palmieri;" pp. 277–282. January, 1866: "Sull' elettrografo atmosferico del Thomson; nota del L Palmieri;" pp. 19–21. April, 1866: "Intorno alla determinazione della vera direzione del vento; nota del L. Palmieri;" pp. 104–105. "Sopra un nuovo metodo ordinato allo studio dell' elettricità atmosferica proposto dal Signor Monnet alla Società delle Scienze Industriali di Lione; nota del L. Palmieri;" pp. 124–125. August, 1866: "Dell' ozono e dell' anti-ozono; nota L. Palmieri;" pp. 268–270. September, 1866: "Sul ricorso delle stelle cadenti nell' Agosto del 1866; nota L Palmieri;" pp. 293–294. "Sulla pressione media del barometro in Napoli, dedotta dalle osservazioni del Prof. Faustino Brioschi; communicazione del Socio Ordinario A. de Gasparis;" pp. 300–304. January, 1867: "Sulla pioggia di stelle cadenti prevista pel 14 Novembre del passato anno; communicazione del L. Palmieri;" p. 18.] (Each monthly number also contains daily notices of the weather and meteorological observations made at Naples.)

Società Reale di Napoli.—Atti dell' Accademia delle Scienze Fisiche e Matematiche. Vol. II. Napoli, 1865. 4to. 602 pp. [Contains: "Sopra un nuovo udometro autegrafico"—2 pp.; "Sull' ozono atmosferico nuove indagini"—16 pp.; "Del periodo diurno dell' elettricità atmosferica e delle sue attenenze con quello delle correnti telluriche"—4 pp.; "Nuovo anemografo elettro-magnetico"—4 pp. These four articles are all by Luigi Palmieri.]

42   METEOROLOGICAL MATERIAL.

Stevens, Linus.—Abstract of rainfall, &c., at Claremont, New Hampshire, for six months ending April 30, 1868.

Stonyhurst Observatory.—Meteorological reports for the years 1860, 1861 1862. 1863, 1864, 1865, 1866, and 1867.

Stumps, S. J.—Meteorological observations made in Hampshire county, West Virginia, from July 1, 1868, to January 1, 1869.

Thornton, Dr.—Summary of meteorological observations taken at Morpeth, New South Wales, during the year 1865.

Trembley, Dr. J. B.—Meteorological synopses for the years 1865, 1866, and 1867, from observations made at Toledo, Ohio. 8vo, 12 pp. each.

United States Naval Observatory.—The November meteors of 1868. 8vo, 10 pp. and chart.

Von Frantzius, Dr. A.—Gaceta Oficial, San José, Costa Rica, containing abstract of meteorology of San José for the year 1867.

Von Oettingen, Dr. Arthur.—Meteorologische Beobachtungen angestellt in Dorpat im Jahre 1867. Redigirt und bearbeitet von Dr. Arthur von Oettingen, Professor der Physik an der Kaiserlichen Universität Dorpat. 8vo. Dorpat, 1868. 118 pp.

Waller, Robert B.—Meteorological observations made at Greensboro', Hale county, Alabama, during the year 1862.

Walton, J. P.—Account of meteors, November 13, 1868. (Newspaper)

Whitcomb, George.—Abstract of meteorological record kept near Charleston Mississippi county, Missouri, from January to September, 1868, inclusive.

Williams, B. C.—Thermometrical record for the month of February, at Ridge farm, Vermillion county, Illinois.

Williams, S. B.—Abstract of observations for each month during 1868, at Lexington, Kentucky.

Wright, Captain J. W. A.—Facts in Meteorology. (Separate paper, published in the Alabama Beacon, Greensboro', Alabama. Newspaper slips.)

Zeledon, José C.—Observaciones meteorológicas hechas en la ciudad de San José (Costa Rica) durante el año de 1868. (Made at the Oficina Central de Estadística and published in the Gaceta Oficial, San José.)

Unknown.—Abstract of meteorological register kept at Fort Yuma, California during the year 1867.

Miscellaneous meteorological notes, Philadelphia.